Singular Spectrum Analysis
of Biomedical Signals

Singular Spectrum Analysis of Biomedical Signals

Saeid Sanei and Hossein Hassani

CRC Press
Taylor & Francis Group
Boca Raton London New York

CRC Press is an imprint of the
Taylor & Francis Group, an **informa** business

CRC Press
Taylor & Francis Group
6000 Broken Sound Parkway NW, Suite 300
Boca Raton, FL 33487-2742

First issued in paperback 2019

© 2016 by Taylor & Francis Group, LLC
CRC Press is an imprint of Taylor & Francis Group, an Informa business

No claim to original U.S. Government works

ISBN-13: 978-1-4665-8927-8 (hbk)
ISBN-13: 978-0-367-37704-5 (pbk)

Visit the Taylor & Francis Web site at
http://www.taylorandfrancis.com

and the CRC Press Web site at
http://www.crcpress.com

Contents

Preface

Recently published literature in biomedical engineering, particularly in analysis of signals and images with a wide range of applications, has attracted more demands for research and development of signal processing tools and algorithms. More recent advancements in signal processing and computerised methods are expected to underpin the key aspects of future progress in biomedical research and technology, particularly on the new measurements and assessment of signals and images from the human body.

Unlike previous publications which focus on the current technology and description of well-established techniques, in this book the aim is to present relatively newly applied concepts for biomedical use, covering a wide range of medical applications.

Although most of the concepts in single channel and multichannel biomedical signal and image processing, particularly the use of adaptive techniques and iterative learning algorithms, have their origin in distinct application areas such as acoustics, biometrics, communications, econometrics, and imaging, it is shown in this book that the new approach, namely singular spectrum analysis (SSA), stems from many recent biological and physiological findings and therefore, it opens a new bio-inspired approach in biomedical engineering.

SSA has its root in subspace analysis and tracking. Given that the recorded biomedical information is often a combination of a number of source signals or signals and artefact, this general approach has always secured a pioneering position in the biomedical signal processing domain.

SSA, on the other hand, works for single channel as well as multichannel data. This advantage explains the exploitation of SSA for a variety of applications such as signal and image restoration, change detection, segmentation, anomaly detection, and prediction, where the recorded signals are mostly single channel.

In addition to the fundamental concepts of SSA published over the past four decades by mathematicians and economists working on financial forecasting, in this book certain new directions in SSA research have been established and adapted to particular biomedical data. A tensor based SSA approach is introduced to analyse multidimensional data, mainly to track the events which migrate from one channel to another channel. Also, SSA has been extended to complex domains in order to exploit the correlation between the two-channel data and work for non-circular complex functions.

Motivated by research in the field of biomedical signal processing over more than two decades, techniques such as genomic data analysis, brain source signal detection and classification, segmentation and anomaly detection of medical images, labelling the stages of sleep from brain signals, restoration of image sequences, prediction of events, and recognition of biometrics are the major focus in this book. In all the cases, comprehensive illustration of new signal processing results in the form of signals, graphs, images, and tables have been provided for a better understanding of the concepts.

Chapter 1 provides a general overview of the major topics in physiological, biological, and biochemical signal processing. It also discusses the concept of biological signals and systems and how the decomposition of signals and extraction of the sources of interest is achievable. Quantification of evolution and prediction of future trends is also discussed.

The SSA concept and its related fundamental details and definitions are the subject of the second chapter of this book. An overview of the main mathematical applications, borrowed from well-established and widely published literature, disregarding our mathematical contributions, is included in this chapter. Most of the concepts and different versions of SSA (univariate and multivariate) discussed in this introductory chapter will be developed and illustrated with various real biomedical signals in the following chapters.

Chapter 3 refers to the concept of processing sleep EEG. The chapter is divided into two main parts, one relating to the physiological aspect of sleep including the stages of sleep, the influence of circadian rhythms, sleep deprivation and psychological effects. Next, the importance of detection and monitoring of brain abnormalities during sleep by electroencephalogram (EEG) analysis is discussed. This leads to an important application of a modified version of SSA, namely tensor SSA. Tensor based SSA, its methodology and a hybrid version, entitled TSSA-EMD is discussed and its application to sleep EEG is illustrated with real data. Flexibility and uniqueness of SSA for solving such issues are also highlighted.

Chapter 4 deals with adaptive SSA and its application. The chapter begins with a description of the adaptive line enhancement (ALE) and SSA based ALE. The concept of incorporating sparsity constraint and its application are also covered. The applications of these techniques are illustrated using an electromyogram (EMG) signal, particularly when corrupted by electrocardiogram (ECG) artefact. This chapter also benefits from the application of adaptive SSA for classification of narrow frequency band system signals. Chapter 4 ends with the application of adaptive filtering to sleep scoring method with experimental results.

The application of SSA based techniques to gait analysis and recognition and the associated subjects with various real examples are covered in Chapter 5. Different techniques, including model free methods are studied. The concepts of Ear-Worn sensor data and its usage for gait analysis as well as the applications of gait analysis in rehabilitation are discussed.

Many signals and system which are complicated in nature are inherently complex. Accordingly, a dedicated chapter of this book is devoted to this important matter. Chapter 6 focuses on complex-valued SSA and its applications to analysis of event-related potentials from EEG. A rigorous mathematical description is made of the theory of the complex-valued matrix, its derivatives, and augmented complex statistics. The augmented complex SSA is studied and its application to improve the extraction of P300 event related potential sub-components is also illustrated.

Chapter 7 is about the change point detection using SSA and its application to eye fundus image analysis. The chapter begins with a brief description of ocular fundus abnormalities and the associated images as well as diabetic retinopathy images. Concise information on most related subjects, including vessel distribution map and vessel segmentation are also given in order to achieve a better understanding of the topic. The main objective of this chapter is to describe how the change point detection capability of SSA enables enhancement or identification of blood vessels and capillaries within eye fundus images.

The subject of prediction of medical and physiological trends is covered in Chapter 8. A number of classical and conventional approaches which have been used for medical prediction are briefly discussed. The application of SSA as a prediction method is illustrated using various examples.

Chapter 9 is devoted to the application of SSA in genetics studies. This chapter also deals with a new approach for the grouping stage of SSA. The link between SSA and Colonial Theory is evaluated, and it is shown how the concept of Colonial Theory enables improving the quality of signal extraction.

In preparation of the content of this book it is assumed that the reader has a background in the fundamentals of digital signal processing and wishes to focus on SSA applications of biomedical signals and images. We hope that the concepts covered in each chapter provide a foundation for future research and development in the field.

In conclusion, we wish to stress that in this book there is no attempt to challenge any clinical or diagnostic knowledge. Instead, the tools and algorithms described in this book can, we believe, enhance the clinically-related information significantly and thereby aid physicians in better diagnosis, treatment and monitoring some of the clinical abnormalities.

Before ending this preface we wish to thank most sincerely our friends and colleagues; Clive Cheongtook, Mansi Ghodsi, Tracey K. M. Lee, and Lilian Tang for their valuable and very useful comments and contributions. Next, my appreciation to our recent PhD students, who contributed to provision of the materials in this book, especially Nader Alharbi, Shirin Enshaeifar, Zara Ghodsi, Xu Huang, Samaneh Kouchaki, Emmanuel Sirimal Silva, and Su Wang.

Saeid Sanei & Hossein Hassani

December 2015

Chapter 1

Introduction

1.1 Physiological, Biological, and Biochemical Processes

More than 250 babies are born every minute, while 150,000 people die daily with population increasing by almost three humans per second. Each of these people lives, thinks, worries, mourns, and daydreams with, and within, a very complex but marvellously organised body. Such a system is influenced by its environment and is able to adapt itself to constructive and destructive waves of changes in a continuously varying earth. To understand the structure and functions of the human body and to understand various biological and biochemical processes involved in human physical and mental development, not only the physiology of the organs but also their functions and perhaps their metabolic activities have to be studied. Most of these studies are done with the help of physiological and biological signs and symptoms measured from the human body invasively or noninvasively, by a variety of sensors and instruments developed through the years. These measurements can be in the form of time series of electric potentials from brain and muscles, images of different modalities, image sequences, heart and lung sounds, blood pressure, eye pressure, body temperature, skin impedance, enzymes, statistics, and many other biometrics. The systems for measuring such information have been improved in recent years, in light of the advances in science and technology. In parallel with the development of new measurement tools and systems, many mathematical and signal processing algorithms have also been introduced and implemented. With the help of such new achievements, diagnosis, treatment, and monitoring of diseases have become much easier, and more lives are saved.

Analysis and modelling of the signal trends, the way they evolve, and the anomalies affecting them have attracted many scientists and researchers around the globe. Biological effects and signals are probably the only type of phenomena which constantly evolve in different aspects (domains). As an example, a concentration of bacteria in one point spreads in different speeds to the nearby cells and tissues depending on many other biochemical, and biological factors. This represents evolution in both time and space. As another example, the electroencephalography (EEG) signals from the medial temporal lobes of epileptic patients can go from chaotic to rhythmic before the seizure onset, and during the ictal period the frequency of the signals decreases slowly. The evolution in such signals is therefore in both time and frequency. In another example, consider the changes in sleep EEG signals [2]. They evolve in their patterns during the four stages of sleep. Understanding and analysis of such patterns reveal much information about human body activities and abnormalities. More importantly, connectivity of the brain lobes varies due to the development of cognitive processes such as mental fatigue, and brain disorders such as dementia or Alzheimer. Processing of the signals from the brain can significantly improve clinical diagnosis and monitoring of the diseases and abnormalities.

Anomalies and so-called novelties in the recorded signals and information are indeed influential when analysing the signal trends and time series. Mutation in genome sequences is a key to the understanding of some life-affecting physiological changes within a live organism. Anomalies in heart beat sequence can be a sign of stroke. Generation of medial temporal spikes can be a sign of seizure development in animals and humans. Many of these signals can be fleeting sources which are transient with no regularities in their occurrence. Consequently, the normal body activity changes due to such anomalies.

Coherency, synchronization, interaction, and stimulation of the human organism involve transmission and processing of multiple sources by a complicated sensory network within the body. Isolation, discrimination, separation, clustering, and classification of these activities are the initial steps towards recognition, modelling, and monitoring of such functionalities. In addition, the activities of the human body change and symptoms evolve due to various cyclic or chaotic natural states of the body such as seizure, the stages of sleep, breathing, heart rate, blood circulation, attention, etc. Characterization, monitoring, quantification, and tracking of such evolution provide an in depth understanding of the human organism.

1.2   Major Topics in Physiological and Biological Signal Processing

As the result of biological, physiological, and biochemical processes in the human body, different types of data can be recorded directly, mostly noninvasively, from the body. These can be divided into three main groups of statistical data representing some biological measures, such as cell counts, medical images such as computerised tomography (CT), magnetic resonance imaging (MRI), positron emission tomography (PET), near infrared spectroscopy (NIRS), and thermography, and finally, signals such as EEG from the brain, electromyography (EMG) from the muscles, ECG from heart, and respiratory signals from lungs. Out of all these modalities, some represent functionality of the brain, heart, muscles, etc. These modalities are more popular in monitoring of patients' conditions.

EEG is a recording modality used to capture the electrical activity in the brain through the electrodes mounted on the scalp. It represents the voltage fluctuations resulting from ionic current flows within the brain neurons. In clinical contexts, EEG refers to the recording of the brain's spontaneous electrical activity over a short period of time from multiple electrodes placed on the scalp. Diagnostic applications generally focus on the spectral content of EEG, that is, the type of neural oscillations that can be observed in EEG signals. In neurology and clinical neuroscience, the main diagnostic application of EEG is in the case of epilepsy, as epileptic activity can create clear abnormalities on a standard EEG study. EEG can be further used in diagnosis of coma, encephalopathies, and brain death. In addition, EEG used to be a first-line method for the diagnosis of tumours, stroke, and other brain disorders, but this use has decreased with the advent of anatomical imaging techniques with much higher spatial resolution such as fMRI. Despite its limited spatial resolution, EEG continues to be a valuable tool for research and diagnosis, especially when millisecond-range temporal resolution (not possible with fMRI) is required, or as very often, when the patient does not have access to other imaging technologies.

Most important derivations of the EEG technique include evoked potentials (EP), which involves averaging the EEG activity time-locked to the presentation of a stimulus of some sort (visual, somatosensory, or auditory). Event related potentials (ERPs) refer to averaged EEG responses that are time-locked to more complex processing of stimuli; this technique is used in understanding of the brain's reactions to various events within the research communities working on cognitive science, cognitive psychology, and psychophysiology.

The brain's electrical charge is maintained by billions of neurons. Neurons are electrically charged (or "polarised") by membrane transport proteins that pump ions across their membranes. Neurons are constantly exchanging ions with the extracellular milieu, for example, to maintain resting potential and to propagate action potentials. Ions of like charge repel each other, and when many ions are pushed out of many neurons at the same time, they can push their neighbors, who push their neighbors, and so on, in a wave. This process is known as volume conduction. This will produce different potential levels over the scalp which can be measured by differential amplifiers. A bank of these amplifiers (up to 250) will then give the signals from each electrode with respect to a common node or versus another electrode.

Obviously, the electric potentials or the corresponding magnetic fields generated by single neurons are far too small to be detected by an EEG or magnetoencephalogram (MEG). EEG activity therefore always reflects the summation of the synchronous activity of thousands or millions of neurons that have similar spatial orientation.

Normal scalp EEG activity shows oscillations at a variety of frequencies. Several of these oscillations have characteristic frequency ranges, spatial distributions and are associated with different states of brain functioning (e.g., waking and, various sleep stages, and mental fatigue). These oscillations represent synchronized activity over a network of neurons. The neuronal networks underlying some of these oscillations are understood (e.g., the thalamocortical resonance underlying sleep spindles), while many others are not (e.g., the system that generates the posterior basic rhythm). Research that measures both EEG and neuron spiking finds the relationship between the two is complex with the power of surface EEG in only two bands (gamma, which is above 20 Hz and delta, which is below 4 Hz) relating to neuron spike activity. The trend and/or synchronization of the rhythms, however, change when the state of the brain varies. Changes in stages of sleep, moving from alert to mental fatigue, transition from preictal state to ictal and then to post ictal or interictal are very good examples of varying the trend of brain rhythms. The changes in connectivity of the brain lobes can also be seen in the cases such as emotions, mental fatigue, Alzheimer, and when the brain is engaged in various mental tasks. A similar non-invasive high temporal resolution technique called MEG measures the magnetic fields generated by neuronal activity of the brain. MEG is a unique and effective diagnostic tool for evaluating brain function in a variety of surgical planning applications. This requires more expensive and less accessible recording equipment. The signals from MEG are often noisier than EEG signals. However, there is a significant advantage of MEG over EEG with regard to the nonlinear effects of various head layers on the EEGs, which do not exist in MEG. Therefore, using this technique, the spatial distributions of the magnetic fields are analysed to localise the sources of the activity within the brain much more accurately than the results that can be achieved using EEG, and the locations of the sources are superimposed on anatomical images, such as MRI, to provide information about both the structure and function of the brain. Like EEG and unlike functional measures such as fMRI, PET, and SPECT, that are secondary measures of brain function reflecting brain metabolism, MEG is a direct measure of brain function. Since the number of magnetic coils used in MEG is often much higher than the number of electrodes in the EEGs, it has higher spatial resolution and often the brain sources can be localised with millimetre precision. The magnetic field passes unaffected through brain tissue and skull. Therefore, it can be recorded outside the head (upper middle). The magnetic field is extremely small, but can be detected by sophisticated sensors that are based on superconductivity. Using this method it is also expected to estimate the intracranial location of the generator sources. A spike in deep brain shows a phase reversal over the left temporal region in the MEG signals. Although MEG is very favourable for brain source localisation, the MEG signal trends can behave like those of an EEG if long term recordings by MEG become possible.

The MEG signals of interest are extremely small, several orders of magnitude smaller than other signals in a typical environment that can obscure the signal. Thus, specialized shielding is required to eliminate the magnetic interference found in a typical urban clinical environment. One significant advantage of EEG over MEG is its simplicity and low price, which allow very long recordings of up to few days from the patients. This cannot be achieved with the other modalities including MEG.

fMRI on the other hand, is an imaging modality meant for the analysis of brain function. The change in the magnetic resonance (MR) signal from neuronal activity is called the hemodynamic response (HDR). It lags behind the neuronal events triggering it by 1 to 2 seconds, since it takes that long for the vascular system to respond to the brain's need for glucose. From this point it typically rises to a peak at about 5 seconds after the stimulus. If the neurons keep firing, say from a continuous stimulus, the peak spreads to a flat plateau while the neurons stay active. After activity stops, the BOLD (blood-oxygenation level dependent) signal falls below the original level, the baseline, causes a phenomenon called the undershoot. Over time the signal recovers to the baseline. The corresponding temporal response signal is called its time-course. The time-course indicates the onset of the events which lead to BOLD construction. For the events which are not time locked, detection of the time-course becomes an important issue. This often requires detection of transient temporal events which may not be possible if only fMRI is used. Detection of BOLD from the brain of an epileptic patient is a good example. To solve this problem a combination of high temporal resolution systems such as EEG and high spatial resolution modality such as fMRI become highly sought. The major problem with these multi-modal data recording systems is the highly destructive effect of the magnetic fields, in the form of gradient and balistocardiogram artefacts over the EEG signals. Also, high resolution fMRI requires a large number of scans during a longer time. The temporal resolution can also be improved by staggering the stimulus presentation across trials. These can lead both to discomfort for the subject inside the scanner and loss of the magnetization signal.

The time resolution needed depends on the brain processing time needed for various events. An example of the broad range here is given by the visual processing system. What the eye sees is registered on the photoreceptors of the retina within a millisecond or so. These signals get to the primary visual cortex via the thalamus in tens of milliseconds. Neuronal activity related to the act of seeing lasts for more than 100 ms. A fast reaction, such as swerving to avoid a car crash, takes around 200 ms. By about half-a-second, awareness and reflection of the incident sets in. Remembering a similar event may take a few seconds, and emotional or physiological changes such as fear arousal may last minutes or hours. Learned changes, such as recognizing faces or scenes, may last days, months, or forever.

A source of nonlinearity in the fMRI response is from the refractory period, where brain activity from a presented stimulus suppresses further activity on a subsequent, similar, stimulus. As stimuli become shorter, the refractory period becomes more noticeable. The refractory period does not change with age, nor do the amplitudes of HDRs. The period differs across brain regions. In both the primary motor cortex and the visual cortex, the HDR amplitude scales linearly with duration of a stimulus or response. In the corresponding secondary regions, the supplementary motor cortex is involved in planning motor behaviur, and the motion-sensitive V5 region, a strong refractory period is seen and the HDR amplitude stays steady across a range of stimulus or response durations. The refractory effect can be used in a way similar to habituation to realize what features of a stimulus a person discriminates as new.

A complex cognitive task may initially trigger high-amplitude signals associated with good performance, but as the subject gets better at it, the amplitude may come down, with the performance staying the same. This is expected to be from the brain more efficiently marshalling neurons to perform the task, decreasing wasteful energy expenditure. The BOLD response across brain regions cannot be compared directly even for the same task, since the density of neurons and the blood-supply characteristics are not constant across the brain. However, the BOLD response can often be compared across subjects for the same brain region and the same task.

Physicians use fMRI to localize the abnormality and assess how risky brain surgery or a similar invasive treatment is for a patient and to learn how a normal, diseased, or injured brain is functioning. They map the brain with fMRI to identify regions linked to critical functions such as speaking, moving, sensing, or planning. This is useful to plan for surgery and radiation therapy of the brain. Clinicians also use fMRI to anatomically map the brain and detect the effects of tumours, stroke, head and brain injury, or diseases such as Alzheimer. Patients with brain pathologies are more difficult to scan with fMRI than are young healthy volunteers, the typical research-subject population. Tumours and lesions can change the blood flow in ways not related to neural activity, masking the neural HDR. Drugs such as antihistamines and even caffeine can affect HDR. Some patients may be suffering from disorders such as compulsive lying, which makes certain studies impossible. Also, since fMRI scanning is a lengthy time process not all the patients can tolerate it. The patients with metallic prosthesis cannot be fMRI scanned either.

The goal of fMRI data analysis is to detect correlations between brain activation and a task the subject performs during the scan. It also aims to discover correlations with the specific cognitive states, such as memory and recognition, induced in the subject. The BOLD signature of activation is relatively weak; however, other sources of noise in the acquired data must be carefully controlled. This means a number of processing steps have to be followed to not only denoise the data but also to detect and recognize the relevant features of the brain activation regions.

ECGs, which noninvasively measure the heartbeat or electrical activity of the heart, can be recorded directly or indirectly from any point in the body where the main blood vessels are closer to the skin. However, to better and more clearly study the heart function a number of electrodes are mounted over the chest or thorax and a few are connected to other body surface points. ECG is therefore used to measure the rate and regularity of heartbeats as well as the size and position of the heart chambers, the presence of any damage to the heart, and the effects of drugs or devices used to regulate the heart (such as a pacemaker). Many heart diseases, particularly problems with heart valves, physical, such as heart stroke, and even mental disorders, such as positive and negative emotional effects and stress, can be diagnosed through ECG records. The ECG pattern is also influenced by metabolic as well as biological changes in the body.

Popular clinical symptoms generally indicating use of electrocardiography however, include cardiac murmurs, syncope or collapse, seizure, perceived cardiac dysrhythmias and myocardial infarction. It is also used to assess patients with systemic disease and to monitor patients during anesthesia as well as critically ill patients.

A typical ECG tracing of the cardiac cycle (heartbeat) consists of a P wave, a QRS complex, a T wave, and a U wave which is normally visible in 50 to 75% of ECGs. The baseline voltage of the ECG is known as the isoelectric line typically measured as the portion of the tracing following the T wave and preceding the next P wave. Figure 1.1 shows a few cycles of a typical ECG waveform.

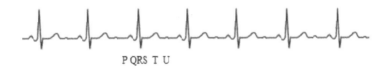

P QRS T U

FIGURE 1.1: A typical ECG waveform.

Another popular biometrics is the record of breathing. Breathing is the process of inhaling and exhaling the air using the lungs. This causes absorption of oxygen by the organs that need oxygen, and the release the carbon dioxide.

The work of breathing is measured as the area of the pleural pressure volume loop. Pleural pressure is assessed using esophageal pressure. The Campbell Diagram [2] shows the different components of the work of breathing. Pressure Time Product (PTP), measures the patient respiratory effort, and has a higher correlation with oxygen consumption than the work of breathing. During assist-controlled ventilation, an increase in flow can decrease the work of breathing by as much as 60% in patients with acute respiratory failure. Higher flow rates can also reduce the inspiratory effort in stable patients with

COPD (chronic obstructive pulmonary disease). During pressure support or assist-controlled ventilation, up to a third of patient effort may fail to trigger the ventilator. Such nontriggering has been shown to result from premature inspiratory efforts that are not sufficient to overcome the elastic recoil associated with dynamic hyperinflation.

Breathing is also affected by many physical, metabolic, and mental effects and the changes are often correlated with heart rate or an onset of abnormal mental activity such as fear.

Another simple, popular, and effective tool for the measurement of blood oxygen is the oximeter. A pulse oximeter (saturometer) is a medical device that indirectly monitors the oxygen saturation of a patient's blood (as opposed to measuring oxygen saturation directly through a blood sample), and changes in blood volume in the skin, producing a photoplethysmogram.

The oximeter sensor is placed on a thin part of the patient's body, usually a fingertip or earlobe, or in the case of an infant, across a foot. Light of two different wavelengths is passed through the patient to a photodetector. The variation in absorbance at each of the two wavelengths is measured, allowing determination of the absorbances due to the pulsing arterial blood alone, excluding venous blood, skin, bone, muscle, fat, and (in most cases) fingernail polish. With NIRS it is possible to measure both oxygenated and deoxygenated hemoglobin on a periperhal scale (possible on both brain and muscle).

Oximeters often have a small display similar to a counter to indicate the oxygen level, so staff can see a patient's oxygenation at all times. Most monitors also display the heart rate. Portable, battery-operated pulse oximeters are also available for home blood-oxygen monitoring. However, currently, most of them can be connected to a PC for analysis and monitoring of the waveforms and statistics.

A pulse oximeter is a simple and useful instrument in any setting where a patient's oxygenation is unstable and needs special care, including hospital intensive care units, surgical operations, recovery, emergency room, and other settings within hospital wards. In addition, it is sometimes used in unpressurized aircraft for the assessment of any patient's oxygenation, and for determining the effectiveness of or need for supplemental oxygen. Assessing a patient's need for oxygen is the most essential element for life; no human life thrives in the absence of oxygen (cellular or gross). Although a pulse oximeter is used to monitor oxygenation, it cannot determine the metabolism of oxygen, or the amount of oxygen being used by a patient. For this purpose, it is necessary to also measure carbon dioxide levels. Pulse oximeter signals can be used not only to determine the oxygen level but also to measure the breathing rate. However, its use in detection of hypoventilation is impaired by the use of supplemental oxygen, as it is only when patients breathe room air that abnormalities in respiratory function can be detected reliably. The output of an oximeter is expected to be a regular periodic signal unless there is instability in the oxygen level or an impairment in breathing. Therefore, it is worth automatically analysing such a signal to monitor the patient's developments.

Another important signal from the body is EMG, which is retrieved both invasively and non-invasively from skeletal muscles. An EMG detects the electrical potential generated by muscle cells when the cells are electrically or neurologically activated. The signals can be analyzed to detect medical abnormalities, activation level, recruitment order or to analyze the biomechanics of human or animal movement. The signals are very noisy and therefore, mathematical and signal processing techniques should be applied to make them useable for diagnosis, and brain-computer interfacing as has been very popular during the past two decades.

There are two kinds of EMG in widespread use: surface EMG and intramuscular (needle and fine-wire) EMG. For intramuscular EMG, a needle electrode or a needle containing two fine-wire electrodes is inserted through the skin into the muscle tissue. The muscle potentials are observed while inserting the electrode. The insertional activity provides valuable information about the state of the muscle and its innervating nerve. Normal muscles at rest make certain, normal electrical signals when the needle is inserted into them. Then the electrical activity when the muscle is at rest is studied. Abnormal spontaneous activity might indicate some nerve and/or muscle damage. Then, the patient is asked to contract the muscle smoothly. The shape, size, and frequency of the resulting motor unit potentials are analyzed. Because the skeletal muscles differ in the inner structure, the electrode has to be placed at various locations to obtain an accurate study.

Intramuscular EMG may be considered too invasive or unnecessary in some cases. Instead, a surface electrode may be used. This technique is used in a number of settings; for example, in a physiotherapy clinic, muscle activation is monitored using a surface EMG and patients have an auditory or visual stimulus to help them know when they are activating the muscle.

This electrophysiologic activity from multiple motor units (neurons) is the signal typically evaluated during an EMG. The composition of the motor unit, the number of muscle fibres per motor unit, the metabolic type of muscle fibres and many other factors affect the shape of the motor unit potentials in the EMG signals.

The study of heart and lung abnormalities moved forward when auscultation and recording of their sounds became possible using electronic stethoscopes. Heart rate is around one beat per second and lung cycles are approximately 15–20 per minute. The nature of sound is different from electric potentials of EEG and EMG and involves much higher frequency bands (similar to speech signal which is often sampled at more than 8 KHz. These patterns are often recorded using a single channel, and it is highly desirable to separate the heart and lung sounds from each other.

Clinical understanding and diagnosis from auscultation requires substantial clinical experience, a fine stethoscope, and good listening skill. Doctors listen to three main organs and organ systems during auscultation: the heart, the lungs, and the gastrointestinal system. When auscultating the heart, clinicians listen for abnormal sounds such as heart murmurs, gallops, and other extra sounds coinciding with the heartbeats. Heart rate is also noted. When listening to lungs, breath sounds such as wheezes, crepitations and crackles are identified. The gastrointestinal system is auscultated to indicate the presence of bowel sounds. Recent electronic stethoscopes can provide noise reduction and signal enhancement. This makes the processing of these signals substantially easier.

1.3   Biological Signals and Systems

Mathematical description of the measurements and the system involved in translating or mapping the biological signals into the sensor signals, particularly in discrete domain, requires fundamental knowledge about signal processing, stochastic processes, and the mathematics involved. Therefore, a number of basic theoretical concepts in relation to those requirements are presented here.

The conventional processes for recording the signals and converting them to digital sample sequences involve two major steps; sampling and quantization. A sampler senses the signals usually in regular time intervals and a quantiser converts the sample amplitudes into the nearest designated quantisation level. The sampling rate should be large enough to comply with the Nyquist rate, i.e. the sampling rate should be bigger than the maximum frequency component of the signals. On the other hand, it should be small enough to meet the hardware and memory requirements of the system. As an example, for sampling the multichannel brain signals the sampling frequencies between 200 Hz and 1000 Hz are very common, whereas for sound recordings the sampling rates of 8 KHz and above are more popular. In order to avoid loss of data, the number of quantization levels is considered reasonably high. In most of the signal acquisition systems 256 levels of quantization are used. This, however, is bounded by the system characteristics and the processor word-length. Combination of sampling rate and quantisation level reveal the system memory requirement since after coding the quantised samples, the overall number of bits correspond to sampling rate, quantisation level, and also the number of channels.

The signals, as time series, have different behaviours and trends. Some have cyclic trends and often the cycle periods remain approximately the same. These signals are deterministic and most of the signal properties are well defined. Some other biological signals are more random with less regularity in their trends. EMG signals look random, with no particular regularities. Seizure signals, on the other hand, are more chaotic long before the seizure onset and turn into ordered signals just before the seizure onset. There are also plenty of transient events within the signals. The event related and movement related cortical potentials are normally infrequent and transient.

Although in some cases multi-modal techniques for data recording, which combines a number of measurement modalities, overcome the limitations of single-modal measurement systems particularly in terms of low time, frequency, or space resolutions, in other cases not more than a single measurement of a single modality system can be used.

1.4 Decomposition of Signals and Extraction of the Sources of Interest

Signals are of different natures. EEGs and MEGs include some of a number of narrow band signal components (semi sinusoids), noise, and also transient event related and movement related signals. To deal with recognition of such signals, source separation and source tracking have been widely used. In addition, the connectivity of the sources as well as the trend of the signal rhythms is subject to change in many cases as described above. For EMG sources the challenge is to detect movement related potentials from very noisy patterns. Adaptive filters, pattern matching, and classification of these patterns have been extensively studied.

Factorization and source separation have been researched in the past two decades mainly for the separation of multichannel data [2]. A priories and any knowledge about the signal sources are exploited in their separation. In general, methods such as time-frequency or time-frequency-space methods can better isolate the events/sources in multi-dimensional spaces. These methods however, do not mimic the actual data features and therefore, are not efficient.

Appreciating the limitation of the number of channels in most of the physiological measurements, singular spectrum analysis (SSA) [4] can decompose the signals to identify the frequency signature of the data over time. On the other hand, it is a subspace approach, since by using it the events can be well defined by a limited number of well distinct orthogonal bases.

1.5 Quantification of Evolution and Prediction of Future Trends

Although for some biomedical signals such as ECG or the oximeter output the normal patterns have regular cycles, the patterns change for abnormalities such as a murmur, collapse of a mitral valve, and hypoxia. For each case the signal trend has to be characterised. This may lead to prediction of the future signal samples, which consequently can be used for diagnosis, monitoring, and alarming the patient.

Patient monitoring and gradual changes of brain activities may be quantified and used as one of the most important indicators of various stages of abnormalities. Quantification of trends and prediction of future events can highly benefit patient monitoring, regulating and minimizing the hospitalization period, decreasing drug administration, and significantly improving the quality of life.

Sleep patterns viewed using the EEG signals change through the sleep stages. Significant variation in normal brain rhythms and the generation of new signal components such as sleep spindles, as discussed in [2], are expected. Analysis of such patterns requires separation of the signals and detection of the desired components to be studied. Later in this book we will see how the stages of sleep can be identified by looking at the changes in alpha and delta (often called slow wave) as the sleep deepens.

Although in many cases the signals are recorded through multichannel sensors, there are many cases where a small number of channels or even only one channel measurement is available. The signals from EMG systems, ECG electrodes, oximeters, stethoscopes for auscultation of heart and lung sounds, and many other recording systems are of this kind.

1.6   Data and Trends; Rhythmic, Noisy, or Chaotic?

Dealing with single channel data, the overarching objective is to decompose the signals as time series into a number of meaningful components. The data often includes one or more desired components and elements of noise of different statistics.

The methods undertaken in assessing the time series analysis should be able to decompose the series into subspaces of different components. SSA is proposed and used here for analysis of single channel data, separating the components, predicting the future trend, and detecting anomalies within the recorded physiological and biological trends.

In order to compensate for the main drawbacks of a basic SSA which is manifested in grouping the eigentriples, some a priori knowledge may be of great help. In addition, the single-channel strategy can be extended to multichannel cases particularly when the system is underdetermined, i.e. the number of measurements is less than the number of sources.

1.7 Generation of Surrogate Data

Generation of surrogate data for mimicking natural events and sources has always been a challenging problem. SSA has great potential for accurate statistical feature extraction. These features can be easily used to generate high quality surrogate data. Such data can be used in modelling, training, and analysis of unknown systems, especially modelling human responses to stimulators. Estimation of future trends and detection of anomalies can then be carried out by assessing the differences between the actual trend and that of the surrogate data.

1.8 Concluding Remarks

In parallel with increasing the number of patients and referrals for clinical examinations on one side and new measurement systems on the other side, the demand for the automation of patient data analysis and even medical diagnosis increases. Single or multichannel information is often available using a variety of measurement systems. SSA is an exceptional mathematical tool which facilitates estimation, detection, prediction, localisation, and classification of desired or undesired events within such data. Similar to some other signal processing techniques, SSA can also be used in data reduction and compression. Multivariate, two-dimentional, tensor-based, complex, quaternion, and constrained SSA are the new complementary approaches which further enhance the SSA applications and extend them to processing of a wider range of information. The following chapters in this book provide some insight into solutions to the above problems.

Bibliography

[1] Sanei, S. (2013). *Adaptive Processing of Brain Signals*. John Wiley & Sons.

[2] Wilfred, C.(1924). Protection of Steam Turbine Disk Wheels from Axial Vibration. *Transactions of the ASME*, 31–160.

[3] Golyandina, N., and Zhigljavsky, A. (2013). *Singular Spectrum Analysis for Time Series*. Springer Science & Business Media.

Chapter 2

Singular Spectrum Analysis

2.1 Introduction

Singular Spectrum Analysis (SSA) has widely and successfully been used in a number of different areas including biomedical signal processing, economics and finance, image processing, earth science and hydrology (for example, see [1]–[16] and references therein). This method is particularly useful for analysing data with complex seasonal patterns and non-stationary trends and particularly in the cases where a single channel measurement is available. The SSA technique is a non-parametric method, and one of its advantages is that it can be used without the need for any assumptions, such as stationarity and normality of the data [1]. Initially, this method decomposes the time series into three components of trend, harmonics and noise. It then reconstructs the series, using the estimated trend and harmonic components and computes the forecasts based on the reconstructed series.

The SSA technique can be applied to a single series or jointly to several series, and in the latter case, by using a multivariate version of that referred to as MSSA. As in the case of parametric modelling, two or more series may be related, which in the context of MSSA has a correspondence in terms of most common contributed components of several series. There are numerous examples of successful applications of the MSSA (for example, see [17]–[27] and references therein).

2.2 Univariate SSA

Let us first start with a brief introduction to the univariate SSA decomposition, reconstruction and forecasting algorithm. In SSA terminology, it is often assumed that the series is noisy, with length N. The SSA technique consists of two complementary stages: decomposition and reconstruction. The noisy series Y_N is first decomposed at the first stage and the noise reduced series is reconstructed at the second stage. Prior to introducing the mathematics involved, here we explain the underlying process in simple terms.

If we assume the simplest structure for a univariate time series $Y_N = (y_1, \ldots, y_N)$ of size N, i.e. that it comprises signal and noise, then Y_N may be represented as:

$$Y_N = \begin{pmatrix} y_1 \\ y_2 \\ \vdots \\ y_N \end{pmatrix} = \begin{pmatrix} s_1 \\ s_2 \\ \vdots \\ s_N \end{pmatrix} + \begin{pmatrix} e_1 \\ e_2 \\ \vdots \\ e_N \end{pmatrix} = S_N + E_N, \qquad (2.1)$$

where S_N represents the signal and E_N the noise.

In order to give clearer indication of the aim of SSA, consider the noisy signal of Figure 2.1. Here, the signal is formed by combination of sine and an exponential trend. SSA decomposes the subspace of the signal into the distinct signal and noise subspaces whereby a clever decision making over the eigenvalue space allows their separation. Accordingly, in this case the SSA technique seeks to extract both the sine and trend components in the time series whilst leaving the noise component out.

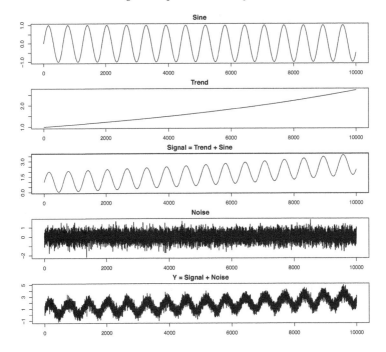

FIGURE 2.1: Sine, trend, signal and noise components in Y.

2.2.1 Stage 1: Decomposition

In the decomposition stage of SSA, the one dimensional time series is segmented into a multidimensional series, by sliding a window of size L over the observed data Y_N of size N, and a matrix is produced by stacking the signal segments which can have $L-1$ overlaps. This procedure is called embedding and results in the so-called trajectory matrix $\mathbf{X}$ with dimensions of L by $K = N - L + 1$.

Step 1: Embedding and Trajectory Matrices

Consider a univariate stochastic process $\{Y_t\}_{t \in \mathbb{Z}}$ and suppose that a realization of size N from this process is available $Y_N = [y_1, y_2, \ldots, y_N]$. The single input at the embedding stage is the SSA choice of L, an integer such that $2 \leq L \leq N - 1$. Embedding can be regarded as a mapping operation that transfers a one-dimensional time series Y_N into a multidimensional series $X_1, \ldots, X_K$ with vectors

$$X_i = [y_i, y_{i+1}, y_{i+2}, \ldots, y_{i+L-1}]^T, \tag{2.2}$$

for $i = 1, 2, \ldots, K$ where $K = N - L + 1$ and T denotes transposition. These vectors group together L time-adjacent observations which are supposed to describe the local state of the underlying process. Vectors X_i are called *L-lagged vectors* (or, simply, *lagged vectors*). The resulting trajectory matrix looks generally like:

$$\mathbf{X} = [X_1,\dots,X_K] = (x_{ij})_{i,j=1}^{L,K} = \begin{pmatrix} y_1 & y_2 & y_3 & \cdots & y_K \\ y_2 & y_3 & y_4 & \cdots & y_{K+1} \\ \vdots & \vdots & \vdots & \ddots & \vdots \\ y_L & y_{L+1} & y_{L+2} & \cdots & y_T \end{pmatrix}. \quad (2.3)$$

Note that the trajectory matrix $\mathbf{X}$ is a Hankel matrix, which means that all the elements along the diagonal $i+j = const$ are equal. Such matrices are sometimes known as persymmetric matrices or, in older literature, orthosymmetric matrices. In general, the Hankel matrix is a matrix whose entries along a parallel to the main anti-diagonal are equal, for each parallel. Equivalently, $\mathbf{X} = (X_{i,j})$ is a Hankel matrix if and only if there exists a sequence $s_1, s_2, \dots$, such that $(x_{i,j}) = s_i + j - 1$, $i, j = 1, 2, \dots$. If s_k are square matrices, then $\mathbf{X}$ is referred to as a block Hankel matrix. Note that we prefer to go forward with matrices as opposed to vectors because matrices enable us to capture more information and they provide more power as we have several vectors instead of one. Moreover, the majority of signal processing techniques can be seen as applied linear algebra and thus we are able to benefit accordingly.

Besides the application in SSA, the trajectory matrix can be used to unify a number of common time series procedures, such as filtering and autoregressive modelling. For example, let β denote any known, fixed $L \times 1$ vector and consider the following:

- For $L = 2$ and $\beta = [-1, 1]$ we can obtain the first differences of the realization as $\beta\mathbf{X}$.

- For any $L \geq 2$ and $\beta = [1/L, 1/L, \dots, 1/L]$ we can obtain a L-order moving average for the realization as $\beta\mathbf{X}$.

- For autoregressive (AR) modelling let Θ denote the parameter vector and Υ denote the vector of innovations. The innovations vector is the difference between the original observed vector and the predicted vector. Write $\mathbf{X}^T\Theta = \Upsilon$ and define the $L \times 1$ and $L \times L$ restriction matrices:

$$R = \begin{bmatrix} 0 \\ 0 \\ \vdots \\ 1 \end{bmatrix}, \quad Q = \begin{bmatrix} I_{L-1} \\ 0_{L-1}^T \end{bmatrix}, \quad (2.4)$$

so that the restricted parameter vector Θ^* is written as $\Theta^* = R - Q\Theta$. Then, the least-squares problem for estimating Θ^* is given by:

$$\min_{\Theta} \Upsilon^T\Upsilon = (R - Q\Theta)^T \mathbf{C_X} (R - Q\Theta), \quad (2.5)$$

with the solution $\hat{\Theta} = (Q^T \mathbf{C_X} Q)^{-1} Q^T \mathbf{C_X} R$ where $\mathbf{C_X} = \mathbf{X}\mathbf{X}^T$.

The window length, L, is an integer between 2 and $N-1$ and is the parameter that should be set in the decomposition stage. The choice of L depends on the structure of the data and no single rule for L covers all the applications. However, in general L should be proportional to the periodicity of the data, large enough to obtain sufficiently separated components but not greater than $N/2$. A complete discussion on the choice of this parameter is given in [1].

Step 2: Singular Value Decomposition

The second step, i.e. the singular value decomposition (SVD) step, decomposes the trajectory matrix $\mathbf{X}$ into its orthogonal bases and represents it as a sum of rank-one bi-orthogonal elementary matrices. Denote $\lambda_1, \ldots, \lambda_L$ as the eigenvalues of $\mathbf{C_X} = \mathbf{X}\mathbf{X}^T$ in decreasing order of magnitude ($\lambda_1 \geq \ldots \lambda_L \geq 0$) and $U_1, \ldots, U_L$ the corresponding eigenvectors.

Set $d = \max(i, \text{such that } \lambda_i > 0) = rank\,\mathbf{X}$. If we denote $V_i = \mathbf{X}^T U_i / \sqrt{\lambda_i}$, then, the SVD of the trajectory matrix can be written as:

$$\mathbf{X} = \mathbf{X}_1 + \cdots + \mathbf{X}_d, \tag{2.6}$$

where $\mathbf{X}_i = \sqrt{\lambda_i} U_i V_i^T$ ($i = 1, \ldots, d$). The matrices $\mathbf{X}_i$ have rank 1. Note that SVD (2.6) is optimal in the sense that among all the matrices $\mathbf{X}^{(r)}$ of rank $r < d$, the matrix $\mathbf{X}^{(r)} = \sum_{i=1}^r X_i$ provides the best approximation to the trajectory matrix $\mathbf{X}$, so that $\| \mathbf{X} - \mathbf{X}^{(r)} \|$ is minimum.

The matrices $\mathbf{X}_i$ are elementary matrices, U_i (in SSA literature they are called 'factor empirical orthogonal functions' or simply EOFs) and V_i (often called 'principal components') stand for the left and right eigenvectors of the trajectory matrix. The collection $(\sqrt{\lambda_i}, U_i, V_i)$ is called the i-th eigentriple of matrix $\mathbf{X}$, $\sqrt{\lambda_i}$ are the singular values and the set of $\{\sqrt{\lambda_i}\}_{i=1}^d$ are often called the spectrum of the matrix $\mathbf{X}$.

Broomhead and King [7] suggested $\frac{1}{L}\mathbf{X}\mathbf{X}^T$, rather than $\mathbf{X}\mathbf{X}^T$, which can be considered for estimation of the lagged-covariance matrix. Vautard and Ghil [18] also proposed the following $L \times L$ Toeplitz matrix $\mathbf{X}$, i.e., a symmetric diagonal-constant matrix with its ij-th element equal to,

$$\frac{1}{N - |i-j|} \sum_{t=1}^{N-|i-j|} y_t y_{t-|i-j|}. \tag{2.7}$$

The SVD of $\mathbf{X}$ is based on the spectral decomposition of the lag-covariance matrix $\mathbf{C_X} \in \mathbb{R}^{L \times L}$. It should be noted that the matrix $\mathbf{C_X}$ is a symmetric and positive semi-definite matrix. Accordingly, it has a complete set of eigenvectors and can be diagonalized in the form $\mathbf{U}\boldsymbol{\Sigma}\mathbf{U}^T$, where $\boldsymbol{\Sigma}$ is the diagonal $L \times L$ matrix of eigenvalues and $\mathbf{U} = (U_1, \ldots, U_L)$ is an orthogonal matrix of eigenvectors of the matrix $\mathbf{C_X}$. The name "singular spectrum analysis" comes from this property of the technique and is a vital component as the entire process is concentrated around obtaining, and analysing this spectrum of singular values to identify and differentiate between the signal and noise in a given time series.

2.2.2   Stage 2: Reconstruction

Step 1: Grouping

At this stage from the pattern of eigenvalues, the desired subspace belonging to the signals of interest and the undesired subspace belonging to noise and artefacts are separated. This often requires some idea about the desired components. The grouping step corresponds to splitting the elementary matrices into several groups and summing the matrices within each group. Let $I = \{i_1, \ldots, i_p\}$ be a group of indices $i_1, \ldots, i_p$. Then the matrix $\mathbf{X}_I$ corresponding to the group I is defined as $\mathbf{X}_I = \mathbf{X}_{i_1} + \cdots + \mathbf{X}_{i_p}$. The split of the set of indices $\{1, \ldots, L\}$ into disjoint subsets $I_1, \ldots, I_m$ corresponds to the representation $\mathbf{X} = \mathbf{X}_{I_1} + \cdots + \mathbf{X}_{I_m}$. The procedure of choosing the sets $I_1, \ldots, I_m$ is called grouping. For a given group I, the contribution of the component $\mathbf{X}_I$ is measured by the share of the corresponding eigenvalues: $\sum_{i \in I} \lambda_i / \sum_{i=1}^{d} \lambda_i$. If the original series contains signal and noise, one then considers two groups of indices, $I_1 = \{1, \ldots, r\}$ and $I_2 = \{r+1, \ldots, L\}$ and associates the group $I = I_1$ with signal component and the group I_2 with noise.

At the grouping step we have the option of analysing the periodogram, scatterplot of right eigenvectors or the eigenvalue functions graph for differentiating between noise and signal. Once we have selected the eigenvalues corresponding to the noise and signal, we can evaluate the effectiveness of this separability via the weighted correlation (w-correlation) statistic. The w-correlation measures the dependence between any two time series and if the separability is sound the two time series will report 0 w-correlation. In contrast, if the w-correlation between the reconstructed components is large, this indicates that the components should be considered as one group [1].

Step 2: Diagonal Averaging

The purpose of diagonal averaging is to transform a matrix to a Hankel matrix which can subsequently be converted to a time series. If z_{ij} stands for an element of a matrix $\mathbf{Z}$, then the k-th term of the resulting series is obtained by averaging z_{ij} over all i, j such that $i + j = k + 1$. By performing the diagonal averaging of all matrix components of $\mathbf{X}_{I_j}$ in the expansion of $\mathbf{X}$ above, we obtain another expansion: $\mathbf{X} = \widetilde{\mathbf{X}}_{I_1} + \ldots + \widetilde{\mathbf{X}}_{I_m}$, where $\widetilde{\mathbf{X}}_{I_j}$ is the diagonalized version of the matrix $\mathbf{X}_{I_j}$. This is equivalent to the decomposition of the initial series $Y_T = (y_1, \ldots, y_T)$ into a sum of m series; $y_t = \sum_{j=1}^{m} \widetilde{y}_t^{(j)}$, where $\widetilde{Y}_T^{(j)} = (\widetilde{y}_1^{(j)}, \ldots, \widetilde{y}_T^{(j)})$ corresponds to the matrix $\widetilde{\mathbf{X}}_{I_j}$.

Note that the SVD of the trajectory matrix $\mathbf{X}$ can be represented as $\mathbf{X} = \sum_{i=1}^{d} \sqrt{\lambda_i} U_i V_i^T = \mathbf{X}_1 + \ldots, \mathbf{X}_d = \sum_{i \in I} \mathbf{X}_i + \sum_{i \notin I} \mathbf{X}_i,$ where $d =$

$\max\{i; i = 1, \ldots, L | \lambda_i > 0\}$ (rank $\mathbf{X} = d$), $V_i = \mathbf{X}^T U_i / \sqrt{\lambda_i}$ ($i = 1, \ldots, d$), $\mathbf{X}_i = \sqrt{\lambda_i} U_i V_i^T$ and $I \subset \{1, \ldots, d\}$. The noise reduced series is reconstructed by $\mathbf{X}_I = \sum_{i \in I} \mathbf{X}_i$ by selecting a set of indices I. In general, however, $\mathbf{X}_I$ does not have a Hankel structure and is not the trajectory matrix of some time series. To overcome this issue, the diagonal averaging over the diagonals $i + j = const$ is used. This corresponds to averaging the matrix elements over the 'antidiagonals' $i + j = k + 1$: the choice $k = 1$ gives $y_1 = y_{1,1}$, for $k = 2$, $y_2 = (y_{1,2} + y_{2,1})/2$, $y_3 = (y_{1,3} + y_{2,2} + y_{3,1})/3$ and so on. Applying diagonal averaging to the matrix $\mathbf{X}_I$ provides a reconstructed signal s_t, and yields the SSA decomposition of the original series y_t as follows $y_t = s_t + \epsilon_t$ ($t = 1, 2, \ldots, N$), where ϵ_t is the residual series after signal extraction. It should be noted that ϵ_t can be reconstructed from diagonal averaging of the matrix $\sum_{i \notin I} \mathbf{X}_i$.

Accordingly, we can summarize the entire SSA process with the aid of the flow chart in Figure 2.2. As appears from Figure 2.2, the algorithm starts with a noisy time series Y_N and the single SSA choice applicable to the Decomposition stage, L, as inputs. Following the embedding procedure the Hankel matrix $\mathbf{X}$ which is then forwarded in the SVD step is obtained. The output from the SVD steps provides us with singular values which are analyzed to identify noise and signal components. Thereafter, the reconstruction stage is reached, whereby the singular values are grouped along with the input of the final SSA choice r which results in the grouping matrices $\mathbf{X}_1, \ldots, \mathbf{X}_L$ as either signal or noise. In the final step we use diagonal averaging to transform the matrices containing signal components into a Hankel matrix.

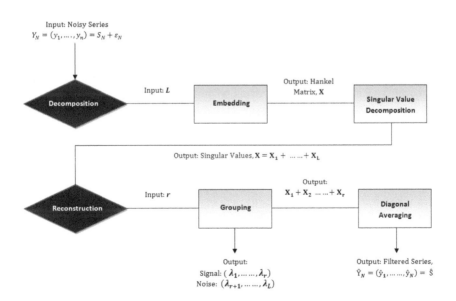

FIGURE 2.2: A summary of the basic SSA process.

It is pertinent to briefly comment on the properties of SSA as identified in the block diagram in Figure 2.2 in relation to its usefulness in application. Consider the decomposition stage up through to the completion of the SVD step. The information attainable here enables obtaining a richer and deeper understanding of the components underlying any given dataset. The reconstruction stage which follows adds further value to the overall SSA process, as it enables one to obtain a new time series according to the objectives underlying the research or analysis. For example, those interested in analysing the trend can attain a time series with the trend alone, whilst the same could be achieved for seasonal variation by considering only the seasonal components. Finally, SSA has the forecasting capability. When coupled with its other properties which gives a whole new sense of flexibility to the modern researchers, the forecasts attainable via SSA itself can be extremely informative and lucrative for analysis and decision making purposes. Forecasts can be tailored to fit the criteria of lowest error (by grouping accordingly) or to predict the future trend or seasonal variation in a given series.

2.2.3   Selection of L and r in Univariate Cases

The SSA technique depends upon two important choices i) the window length L, and ii) the effective r. The choice of L depends on several criteria including complexity of the data, the aim of the analysis and in the context of prediction, the forecasting horizon. Thus, the improper choice of L would imply an inferior decomposition, and then inaccurate forecasting results. On the other hand, by selecting r smaller than the true number of eigenvalues, some parts of the signal(s) will be lost, and then the reconstructed series becomes less accurate. However, if one takes r greater than the value that it should be, then noise components will be included in the reconstructed series. The situation is even more vague in the case of MSSA in which the similarity and orthogonality among series play an important rule for selecting L and r. Moreover, the multivariate case deals with a block trajectory Hankel matrix with special features rather than one Hankel matrix. This makes the problem even more complex.

Furthermore, in order to find the optimal value of L, we need to investigate the concept of separability, which is one of the main methodological concepts in SSA and characterizes how well different components can be separated from each other. It should be noted that, the SSA decomposition stage provides more accurate results if the resulting additive components of the series are approximately separable from each other.

The first step of the SSA algorithm provides a Hankel trajectory matrix, which plays an important role in the SSA technique, as the other steps depend on its structure and the extracted eigenvalues obtained from this matrix [28]. In MSSA this matrix can be formed differently. This will change the forecasting results significantly in some situations. Furthermore, depending on the structure of this matrix different forecasting algorithms are formed.

This matrix also depends on the window length L. Here, the characteristics of the trajectory matrix with respect to different values of the L is studied. Note also that the second SSA choice, which is the number of eigenvalues r, is also very important for the reconstruction stage, which will provide the noise free series, and the forecasting purpose.

The selection of r for the univariate case is still an open problem. However, there are several recommendations for some particular types of series, like the seasonal time series (see, for example, [1] and [4]). As demonstrated in some later chapters, particularly for the design of adaptive filters, the corresponding eigenvalues can be selected adaptively. This issue is even more difficult for the multivariate case, as each eigenvalue contains information about all the series considered in the multivariate analysis. Here, the optimality aspect of the SSA choices is considered for the multivariate case and various bounds for r are introduced. Similar to other related multivariate analysis, it is also important to know how different time series change in relation to each other, or how they are associated. To evaluate this, the trajectory matrix, which is a block Hankel matrix, needs to be considered. For the SSA analysis, orthogonality among the reconstructed series, that are obtained from the extracted eigenvalues from the trajectory matrix, is foremost. Therefore, the number of common/matched components among time series needs to be considered. Here, we evaluate the effect of most common contributed components among series from a theoretical aspects and also through comprehensive simulation studies. How similarity and dissimilarity among series affect the forecasting performance is also evaluated.

The use of ancillary and supplementary evidences in real life circumstances raises the performance of the signal processing technique to a certain extent. In SSA analysis, the auxiliary information about the signal aids in choosing L and r. For instance, the presence of an annual periodicity in the series advises that the frequency $k/12$ ($k = 1, ..., 12$) should be taken into account for selection of L and r. In brief, it is possible to identify four distinct methods for determining the SSA choices of L and r, i.e. analysis of singular values, pairwise scatterplots, the periodogram, and the concept of separability [1]. These are explained in detail below.

2.2.3.1 Singular Values

Usually every harmonic component with a different frequency produces two eigentriples with close singular values (except for frequency 0.5 which provides one eigentriple with a saw-tooth singular vector). It will be clearer if T, L and K are sufficiently large.

Another useful insight is provided by checking breaks in the eigenvalue spectra. As a rule, a pure noise series produces a slowly decreasing sequence of singular values.

Therefore, explicit plateaux in the eigenvalue spectra prompt the ordinal numbers of the paired eigentriples. Figure 2.3 depicts the plot of the logarithms of the 24 singular values for a real series.

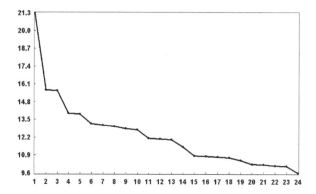

FIGURE 2.3: Logarithms of the 24 eigenvalues.

Five evident pairs with almost equal leading singular values, correspond to five (almost) harmonic components of the real time series: eigentriple pairs 2–3, 4–5, 7–8, 9–10 and 11–12 are related to harmonics with specific periods (we show later that they corresponds to periods 12, 6, 2.5, 4 and 3).

2.2.3.2 Pairwise Scatterplots

In practice, the singular values of the two eigentriples of a harmonic series are often very close to each other, and this fact simplifies the visual identification of the harmonic components. An analysis of the pairwise scatterplots of the singular vectors allows one to visually identify those eigentriples that correspond to the harmonic components of the series, provided these components are separable from the residual component.

Consider a pure harmonic with a frequency w, a certain phase, amplitude and ideal situation where $P = 1/w$ is a divisor of the window length L and K. Since P is an integer, it is a period of the harmonic. In the ideal situation, the left eigenvectors and principal components have the form of sine and cosine sequences with the same P and the same phase. Thus, the identification of the components that are generated by a harmonic is reduced to the determination of these pairs.

The pure sines and cosines with equal frequencies, amplitudes, and phases create the scatterplot with the points lying on a circle. If $P = 1/w$ is an integer, then these points are the vertices of the regular P-vertex polygon. For the rational frequency $w = m/n < 0.5$ with relatively prime integers m and n, the points are the vertices of the scatterplots of the regular n-vertex polygon. Figure 2.4 depicts scatterplots of the 6 pairs of sine/cose sequence (without noise) with zero phase, the same amplitude and periods 12, 6, 4, 3, 2.5 and 2.4.

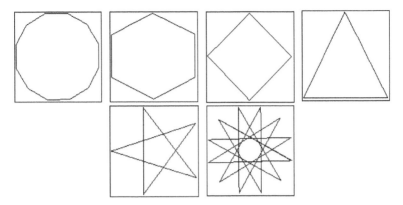

FIGURE 2.4: Scatterplots of the 6 pairs of sines/cosines.

2.2.3.3   Periodogram Analysis

The periodogram analysis of the original series and eigenvectors may help in making the proper grouping; it tells us which frequency must be considered. We must then look for the eigentriples whose frequencies coincide with the frequencies of the original series.

If the periodograms of the eigenvector have sharp sparks around some frequencies, then the corresponding eigentriples must be regarded as those related to the signal component.

Figure 2.5 depicts the periodogram of the paired eigentriples (2–3, 4–5, 7–8, 9–10, 11–12). The information arising from Figure 2.5 confirms that the above mentioned eigentriples correspond to the periods $12, 6, 2.5, 4$ and 3 which must be regarded as selected eigentriples in the grouping step with another eigentriple we need to reconstruct the series.

2.2.3.4   Separability

The main concept in studying SSA properties is 'separability', which characterizes how well different components can be separated from each other. SSA decomposition of the series Y_T can only be successful if the resulting additive components of the series are approximately separable from each other. The following quantity (called the weighted correlation or *w-correlation*) is a natural measure of dependence between two series $Y_T^{(1)}$ and $Y_T^{(2)}$:

$$\rho_{12}^{(w)} = \frac{\left(Y_T^{(1)}, Y_T^{(2)}\right)_w}{\parallel Y_T^{(1)} \parallel_w \parallel Y_T^{(2)} \parallel_w}$$

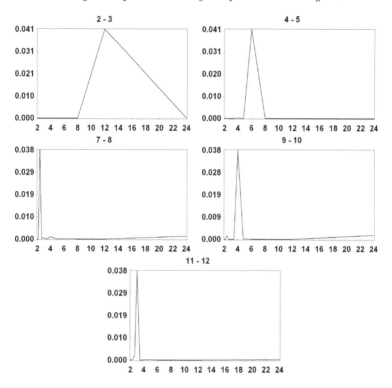

FIGURE 2.5: Periodograms of the paired eigentriples (2–3, 4–5, 7–8, 9–10, 11–12).

where $\| Y_T^{(i)} \|_w = \sqrt{\left(Y_T^{(i)}, Y_T^{(i)} \right)_w}$, $\left(Y_T^{(i)}, Y_T^{(j)} \right)_w = \sum_{k=1}^{T} w_k y_k^{(i)} y_k^{(j)}$, $(i, j = 1, 2)$

$w_k = \min\{k, L, T - k\}$ (here we assume $L \leq T/2$).

If the absolute value of the w-correlations is small, then the corresponding series are almost w-orthogonal, but, if it is large, then the two series are far from being w-orthogonal and are therefore badly separable. So, if two reconstructed components have zero w-correlation it means that these two components are separable. Large values of w-correlations between reconstructed components indicate that the components should possibly be gathered into one group and correspond to the same component in SSA decomposition.

Figure 2.6 shows the w-correlations for the 24 reconstructed components in a 20-grade grey scale from white to black corresponding to the absolute values of correlations from 0 to 1.

In the conclusion of this section it is clear that SSA also provides added flexibility in terms of selection of the SSA choices of L and r. Once again, depending on the objective of the analysis, researchers could consider any of the aforementioned approaches to determine the best SSA choices to model

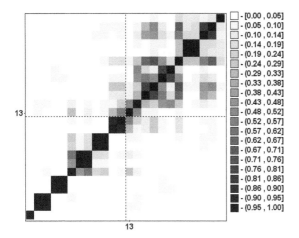

FIGURE 2.6: Matrix of w-correlations for the 24 reconstructed components.

their data. Analysis of the periodogram for any given series can be a sound starting point for determining which L to use. Thereafter, in terms of selecting r, those interested in capturing the leading information and paired eigenvectors in a given dataset can analyse the logarithms of singular values and group accordingly. On the other hand, if certain seasonal components were only of interest, an analysis of the pairwise scatterplots could help with the determination of r. Moreover, if obtaining optimal separability between signal and noise components is crucial for the underlying study, researchers could then turn to the w-correlation matrix for this purpose. In what follows the SSA forecasting algorithms are introduced.

2.2.4   SSA Forecasting Algorithm

An important application of SSA is in forecasting either the individual components of the series and/or the reconstructed series itself. The SSA technique can be applied in forecasting any time series that approximately satisfies the linear recurrent formulae (LRF):

$$y_j = \sum_{i=1}^{L-1} \alpha_i y_{j-i}, \qquad L \leq j \leq N \qquad (2.8)$$

where coefficients $\alpha_1, \ldots, \alpha_d$ are achieved based on U_i. An important property of the SSA decomposition is that, if the original time series Y_T satisfies the LRF (2.8), then for any N and L there are at most d nonzero singular values in the SVD of the trajectory matrix $\mathbf{X}$; therefore, even if the window length L and $K = N - L + 1$ are larger than d, we only need at most d matrices $\mathbf{X}_i$ to reconstruct the series. Let us now formally describe the algorithm for the SSA forecasting method.

The reconstructed series $\widetilde{Y}_N = (\widetilde{y}_1, \ldots, \widetilde{y}_N)$ is then used for forecasting new data points, say y_{N+1}. The value of y_{N+1} is forecasted in such a way that the vector $Y_{N+1} = (\widetilde{y}_{K+1}, \ldots, \widetilde{y}_N, y_{N+1})$ is nearest to the subspace of $\mathbb{R}^L$ spanned by the eigenvectors U_j, $j \in I$. If one considers the Euclidean norm in $\mathbb{R}^L$ then the problem is simplified as the problem of minimising the norm of the difference vector between Y_{N+1} and its orthogonal projection $\mathbf{P}Y_{N+1}$

$$\| (\mathbf{I} - \mathbf{P})Y_{N+1} \|^2 \to \min$$

varying y_{N+1}, where $\mathbf{P} = \sum_{i \in I} U_i U_i^T$ and $\mathbf{I}$ is an identity matrix (note also that $\mathbf{PX} = \mathbf{X}_I$). This results in

$$0 = \frac{d}{dy_{T+1}} \| (\mathbf{I} - \mathbf{P})Y_{T+1} \|^2 = 2Y_{T+1}^T (\mathbf{I} - \mathbf{P})(\mathbf{I} - \mathbf{P})^{\triangleleft}$$

where $(\mathbf{I} - \mathbf{P})^{\triangleleft}$ denotes the last column of matrix $(\mathbf{I} - \mathbf{P})$. Thus, minimizing vector Y_{N+1} is orthogonal to $(\mathbf{I} - \mathbf{P})(\mathbf{I} - \mathbf{P})^{\triangleleft}$ and the recurrence formula to forecast the new data point y_{N+1} is as follows:

$$y_{N+1} = -\frac{1}{\alpha_L} \sum_{i=1}^{L-1} \alpha_i s_{N+1-L+i}, \qquad \alpha_L \neq 0$$

The above forecasting method is called the recurrent approach and can be extended for h-step ahead forecast. Now formally consider the SSA forecasting algorithm, as proposed in [1].

1. Consider a time series $Y_N = (y_1, \ldots, y_N)$ with length N.

2. Fix the window length L.

3. Consider the linear space $\mathcal{L}_r \subset \mathbf{R}^L$ of dimension $r < L$. It is assumed that $e_L \notin \mathcal{L}_r$, where $e_L = (0, 0, \ldots, 1) \in \mathbf{R}^L$.

4. Construct the trajectory matrix $\mathbf{X} = [X_1, \ldots, X_K]$ of the time series Y_N.

5. Construct the vectors U_i $(i = 1, \ldots, r)$ from the SVD of $\mathbf{X}$. Note that U_i is an orthonormal basis in $\mathcal{L}_r$.

6. Orthogonal projection step; compute matrix $\widehat{\mathbf{X}} = [\widehat{X}_1 : \ldots : \widehat{X}_K] = \sum_{i=1}^{r} U_i U_i^T \mathbf{X}$. The vector $\widehat{X}_i$ is the orthogonal projection of X_i onto the space $\mathcal{L}_r$.

7. Hankellization step; construct the matrix $\widetilde{\mathbf{X}} = \mathcal{H}\widehat{\mathbf{X}} = [\widetilde{X}_1 : \ldots : \widetilde{X}_K]$.

8. Set $v^2 = \pi_1^2 + \ldots + \pi_r^2$, where π_i is the last component of the vector U_i $(i = 1, \ldots, r)$. Moreover, assume that $e_L \notin \mathcal{L}_r$. This implies that $\mathcal{L}_r$ is not a vertical space. Therefore, $v^2 < 1$.

9. Determine vector $A = (\alpha_1, \ldots, \alpha_{L-1})$:

$$A = \frac{1}{1 - v^2} \sum_{i=1}^{r} \pi_i U_i^{\nabla}.$$

where $U^{\nabla} \in \mathbf{R}^{L-1}$ is the vector consisting of the first $L - 1$ components of the vector $U \in \mathbf{R}^L$. It can be proved that the last component y_L of any vector $Y_L = (y_1, \ldots, y_L)^T \in \mathfrak{L}_r$ is a linear combination of the first y_{L-1} components, i.e.

$$y_L = \alpha_1 y_{L-1} + \ldots + \alpha_{L-1} y_1,$$

and this does not depend on the choice of a basis $U_1, \ldots, U_r$ in the linear space $\mathfrak{L}_r$.

10. Define the time series $Y_{N+h} = (y_1, \ldots, y_{N+h})$ by the formula

$$y_i = \begin{cases} \widetilde{y}_i & \text{for } i = 1, \ldots, N \\ \sum_{j=1}^{L-1} \alpha_j y_{i-j} & \text{for } i = N+1, \ldots, N+h \end{cases} \tag{2.9}$$

where $\widetilde{y}_i$ $(i = 1, \ldots, N)$ are the reconstructed series. Then, $y_{N+1}, \ldots, y_{N+h}$ are the h step ahead recurrent forecasts.

The above SSA forecasting algorithm is called the recurrent forecasting approach (RSSA). An alternative approach is called the Vector SSA forecasting algorithm (VSSA). In the RSSA approach we consider the last component of the reconstructed vector for forecasting, whilst in the VSSA the entire eigenvector is considered for estimating the new values, and this is in fact the main difference between RSSA and VSSA. The VSSA approach is as follows. Consider the following matrix:

$$\Pi = \mathbf{V}^{\nabla} (\mathbf{V}^{\nabla})^T + (1 - v^2) A A^T \tag{2.10}$$

where $\mathbf{V}^{\nabla} = [U_1^{\nabla}, \ldots, U_r^{\nabla}]$. Now consider the linear operator

$$\theta^{(v)} : \mathfrak{L}_r \mapsto \mathbf{R}^L \tag{2.11}$$

where

$$\theta^{(v)} U = \begin{pmatrix} \Pi U^{\nabla} \\ A^T U^{\nabla} \end{pmatrix}. \tag{2.12}$$

Define vector Z_i as follows:

$$Z_i = \begin{cases} \widetilde{X}_i & \text{for } i = 1, \ldots, K \\ \theta^{(v)} Z_{i-1} & \text{for } i = K+1, \ldots, K+h+L-1 \end{cases} \tag{2.13}$$

where, $\widetilde{X}_i$'s are the reconstructed columns of the trajectory matrix after grouping and eliminating noise components. Now, by constructing matrix $\mathbf{Z} = [Z_1, \ldots, Z_{K+h+L-1}]$ and performing diagonal averaging we obtain a new series $y_1, \ldots, y_{N+h+L-1}$, where $y_{N+1}, \ldots, y_{N+h}$ form the h terms of the SSA vector forecast.

2.3   Multivariate SSA

The univariate SSA forecasting algorithm is based on two main forecasting approaches; recurrent and vector. However, for the multivariate case only the recurrent approach has been developed. The discrepancy between the recurrent and vector approaches in the multivariate SSA (MSSA) is mainly due to organizing the single trajectory matrix $\mathbf{X}$ of each series into the block trajectory matrix in the multivariate case. Two trajectory matrices can be organized either in vertical or horizontal forms. The MSSA with vertical form shall be called VMSSA and with horizontal form, HMSSA[42]. In general there are two forms of the block trajectory matrix in MSSA along with two forecasting procedures. Consequently, considering the recurrent and vector approaches, there are four different forecasting algorithms as follows:

$$
\text{MSSA forecasting approach} = \begin{cases} \text{HMSSA} & \begin{cases} \text{Recurrent approach} \\ \text{Vector approach} \end{cases} \\ \\ \text{VMSSA} & \begin{cases} \text{Recurrent approach} \\ \text{Vector approach} \end{cases} \end{cases}
$$

2.3.1   MSSA: Vertical Form

The MSSA Vertical form is referred to as VMSSA. Consider M time series with different series length N_i; $Y_{N_i}^{(i)} = (y_1^{(i)}, \ldots, y_{N_i}^{(i)})$ $(i = 1, \ldots, M)$. Here, we provide the procedure for the multivariate case only, the univariate form can be acquired by setting $M = 1$ for all the multivariate algorithms considered in this chapter.

Stage I. Decomposition

Step 1: Embedding. Embedding is a mapping that transfers a one-dimensional time series $Y_{N_i}^{(i)} = (y_1^{(i)}, \ldots, y_{N_i}^{(i)})$ into a multidimensional matrix $[X_1^{(i)}, \ldots, X_{K_i}^{(i)}]$ with vectors $X_j^{(i)} = (y_j^{(i)}, \ldots, y_{j+L_i+1}^{(i)})^T \in \mathbf{R}^{L_i}$, where L_i $(2 \leq L_i \leq N_i - 1)$ is the window length for each series with length N_i and $K_i = N_i - L_i + 1$. The result of this step is the trajectory matrix $\mathbf{X}^{(i)} = [X_1^{(i)}, \ldots, X_{K_i}^{(i)}] = (x_{mn})_{m,n=1}^{L_i, K_i}$. As it was mentioned above, the trajectory matrix $\mathbf{X}^{(i)}$ is a Hankel matrix. Thus, the above procedure for each series separately provides M different $L_i \times K_i$ trajectory matrices $\mathbf{X}^{(i)}$ $(i = 1, \ldots, M)$. To form a new block Hankel matrix in a vertical form, it is required to have $K_1 = \ldots, K_M = K$. Accordingly, this version of MSSA enables us to have various window length L_i and different series length N_i, but similar K_i for all series. The result of this step is the following block Hankel trajectory matrix

$$\mathbf{X}_V = \begin{bmatrix} \mathbf{X}^{(1)} \\ \vdots \\ \mathbf{X}^{(M)} \end{bmatrix},$$

where, $\mathbf{X}_V$ indicates that the output of the first step is a block Hankel trajectory matrix formed in a *vertical* form.

Step 2: SVD. In this step, we perform the SVD of $\mathbf{X}_V$. Denote $\lambda_{V_1}, \ldots, \lambda_{V_{L_{sum}}}$ as the eigenvalues of $\mathbf{X}_V \mathbf{X}_V^T$, arranged in decreasing order $\left(\lambda_{V_1} \geq \ldots \lambda_{V_{L_{sum}}} \geq 0 \right)$ and $U_{V_1}, \ldots, U_{V_{L_{sum}}}$, the corresponding eigenvectors, where $L_{sum} = \sum_{i=1}^{M} L_i$. Note also that the structure of the matrix $\mathbf{X}_V \mathbf{X}_V^T$ is as follows:

$$\mathbf{X}_V \mathbf{X}_V^T = \begin{bmatrix} \mathbf{X}^{(1)}\mathbf{X}^{(1)T} & \mathbf{X}^{(1)}\mathbf{X}^{(2)T} & \cdots & \mathbf{X}^{(1)}\mathbf{X}^{(M)T} \\ \mathbf{X}^{(2)}\mathbf{X}^{(1)T} & \mathbf{X}^{(2)}\mathbf{X}^{(2)T} & \cdots & \mathbf{X}^{(2)}\mathbf{X}^{(M)T} \\ \vdots & \vdots & \ddots & \vdots \\ \mathbf{X}^{(M)}\mathbf{X}^{(1)T} & \mathbf{X}^{(M)}\mathbf{X}^{(2)T} & \cdots & \mathbf{X}^{(M)}\mathbf{X}^{(M)T} \end{bmatrix}. \tag{2.14}$$

The structure of the matrix $\mathbf{X}_V \mathbf{X}_V^T$ is similar to the covariance matrix in the classical multivariate statistical analysis literature. The matrix $\mathbf{X}^{(i)}\mathbf{X}^{(i)T}$, which is used in the univariate SSA, for the series $Y_{N_i}^{(i)}$, appear along the main diagonal and the products of two Hankel matrices $\mathbf{X}^{(i)}\mathbf{X}^{(j)T}$ ($i \neq j$), which are related to the series $Y_{N_i}^{(i)}$ and $Y_{N_j}^{(j)}$, appears in the off-diagonal. The SVD of $\mathbf{X}_V$ can be written as $\mathbf{X}_V = \mathbf{X}_{V_1} + \cdots + \mathbf{X}_{V_{L_{sum}}}$, where $\mathbf{X}_{V_i} = \sqrt{\lambda_i} U_{V_i} V_{V_i}{}^T$ and $V_{V_i} = \mathbf{X}_V^T U_{V_i}/\sqrt{\lambda_{V_i}}$ ($\mathbf{X}_{V_i} = \mathbf{0}$ if $\lambda_{V_i} = 0$).

Stage II. Reconstruction

Both grouping and diagonal averaging are performed similar to univariate SSA.

2.3.2 MSSA: Horizontal Form

The MSSA Horizontal form is referred to as HMSSA. The decomposition and reconstruction stages of the HMSSA algorithm are similar to those provided above for VMSSA except for the structure of the block Hankel matrix. Let us have M different $L_i \times K_i$ trajectory matrices $\mathbf{X}^{(i)}$ ($i = 1, \ldots, M$). To construct a block Hankel matrix in the horizontal form we need to have $L_1 = L_2 = \ldots = L_M = L$. Therefore, we have different values of K_i and series length N_i, but similar L_i. The result of this step is as follows:

$$\mathbf{X}_H = [\ \mathbf{X}^{(1)} : \ \ \mathbf{X}^{(2)} : \ \ \cdots \ \ : \mathbf{X}^{(M)}\ ].$$

Hence, the structure of the matrix $\mathbf{X}_H \mathbf{X}_H^T$ is as follows:

$$\mathbf{X}_H \mathbf{X}_H^T = \mathbf{X}^{(1)} \mathbf{X}^{(1)^T} + \cdots + \mathbf{X}^{(M)} \mathbf{X}^{(M)^T}. \tag{2.15}$$

As it appears in the structure of the matrix $\mathbf{X}_H \mathbf{X}_H^T$ in HMSSA, we do not have any cross-product between Hankel matrices $\mathbf{X}^{(i)}$ and $\mathbf{X}^{(j)}$. Moreover, in this format, the sum of $\mathbf{X}^{(i)} \mathbf{X}^{(i)^T}$ provides the new block Hankel matrix. Note also that performing the SVD of $\mathbf{X}_H$ in HMSSA yields L eigenvalues as the univariate SSA, whilst we have $L_{sum} = \sum_{i=1}^{M} L_i$ eigenvalues in VMSSA.

In the h-step ahead forecasts of each series in the HMSSA approach are achieved by the same LRF generated considering all series in a multivariate system. Table 2.1 provides a general view on the similarities and dissimilarities of VMSSA and HMSSA forecasting algorithms from different perspectives; the series length, the value of the window length (L_i), the number of non-zero singular values obtained from the block trajectory matrix and the LRF.

TABLE 2.1: Similarities and dissimilarities between the VMSSA and HMSSA recurrent forecasting algorithms.

Method	Series Length	L_i	K_i	Number of λ_i	LRF
VMSSA-R	Different	Different	Equal	$\sum L_i$	Different
HMSSA-R	Different	Equal	Different	L	Equal

As appears from Table 2.1 there are some restrictions on selecting of the values of L and K. These restrictions can be avoided by considering the following arrangement:

$$\mathbf{X}_D = \begin{bmatrix} \mathbf{X}^{(1)} & \mathbf{0} & \cdots & \mathbf{0} \\ \mathbf{0} & \mathbf{X}^{(2)} & \cdots & \mathbf{0} \\ \vdots & \vdots & \ddots & \vdots \\ \mathbf{0} & \mathbf{0} & \cdots & \mathbf{X}^{(M)} \end{bmatrix}. \tag{2.16}$$

Certainly matrix $\mathbf{X}_D$ has more dimensions than matrices $\mathbf{X}_V$ and $\mathbf{X}_H$; however, this gives us more freedom in choosing L and K. The new arrangement might be useful in the noise reduction stage as we can increase L_i for each series, to the maximum value which is $N_i/2$ and also for Hankelization purposes using an iterative approach rather than using simple averaging [42].

There also exist the HMSSA and VMSSA vector forecasting algorithms [42]. The MSSA vector forecasting algorithm has similar inputs as the MSSA recurrent forecasting algorithm. The mathematical basis of both forecasting algorithms with theoretical background and practical applications has been extensively studied in [42]. The idea of the vector forecasting approach is based on the continuation of the columns of the trajectory matrix. To perform the vector forecasting algorithm, the new vector must satisfy the following i) belong to the column space of the eigenvectors matrix of the trajectory matrix, and ii) the newly constructed matrix obtained by adding these vectors should be approximately a Hankel matrix. The continuation vectors can be obtained by an operator that has two parts; the perpendicular projection operator on the column space of the eigenvectors matrix of the trajectory matrix, and the linear recurrent coefficients vector.

2.4 Optimal Values of L and r in MSSA

2.4.1 L

The usual bound for L in the univariate SSA is $2 \leq L \leq N - 1$ and its optimal value, at least for the reconstruction stage, is close to Median$\{1, \ldots, N\}$ [31]. It should be mentioned that this bound may not provide the optimal forecasting results (see, for example, [27]). Note also that the refined series, achieved from the reconstruction stage, using the trajectory matrix $\mathbf{X}_{L \times K}$ is similar if we use $\mathbf{X}_{L \times K}^{T} = \mathbf{X}_{K \times L}$. Therefore, considering L more than Median$\{1, \ldots, N\}$ provides similar results in the univariate case.

However, the optimal value for the multivariate case is different. The optimal value of the multivariate case has been introduced in [42]. A summary of main important results is briefly discussed below. Note that if $\mathbf{X}$ with dimension $LM \times K$ denotes the trajectory matrix of VMSSA, then $\mathbf{X}^{T}$ is the trajectory matrix of HMSSA with dimension $K \times ML$ (assuming all the series have similar series length and similar L).

Let d denote the rank of the trajectory matrix $\mathbf{X}$ in multivariate SSA (either VMSSA or HMSSA). Thus, the maximum rank of the trajectory matrix $\mathbf{X}$ is achieved at $L = \left\lceil \frac{1}{M+1}(N+1) \right\rceil$ and $L = \left\lceil \frac{M}{M+1}(N+1) \right\rceil$ for VMSSA and HMSSA, respectively, where $\lceil u \rceil$ denotes the closest integer number to u [42].

This indicates that the maximum rank of $\mathbf{X}$ is attained when $|L(M+1) - (N+1)|$ is minimised. Thus, $L = \left\lceil \frac{N+1}{M+1} \right\rceil$. Similarly, it is straightforward to prove that $L = \left\lceil \frac{M}{M+1}(N+1) \right\rceil$ provides the optimal L for VMSSA. Moreover, these values of L are the first $(M+1)$-quantile and the Mth $(M+1)$-quantile of the integer set $\{1, \ldots, N\}$. Therefore, the optimal values of L in the reconstruction stage of MSSA, having M series with length N, are as follows:

$$L = \begin{cases} \left\lceil (\frac{1}{M+1})(N+1) \right\rceil & \text{VMSSA}; \\[2ex] \left\lceil (\frac{M}{M+1})(N+1) \right\rceil & \text{HMSSA}. \end{cases} \tag{2.17}$$

Figure 2.7 shows the value of d, for two series with equal lengths $N = 50$, with respect to different values of L. As it appears from Figure 2.7 in the first case ($M = 2$), the rank of the trajectory matrix is maximised at $L = \left\lceil \frac{50+1}{2+1} \right\rceil = 17$, and $L = \left\lceil \frac{2}{3}(50+1) \right\rceil = 34$ for VMSSA and HMSSA, respectively. Moreover, in the second case ($M = 3$), the maximum values are $L = 13$ and 38 for VMSSA and HMSSA, respectively. Thus, the results confirm that choosing quantiles of $\{1, \ldots, N\}$ provides the maximum rank of the trajectory matrix $\mathbf{X}$ for both VMSSA and HMSSA.

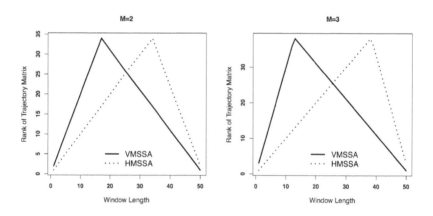

FIGURE 2.7: The rank of the trajectory matrix in VMSSA and HMSSA, for two series with equal lengths $N = 50$, with respect to different values of L and M [42].

2.4.2　*r*

Let us first consider the issue of orthogonality and similarity (with respect to some criteria) between two matrices $\mathbf{X}^{(i)}$ and $\mathbf{X}^{(j)}$ ($i \neq j$) which plays an important role in the MSSA analysis. Consider two trajectory Hankel matrices $\mathbf{X}^{(i)}$ and $\mathbf{X}^{(j)}$ ($i \neq j$) corresponding to the series $Y_N^{(i)}$ and $Y_N^{(j)}$, respectively. Then, matrices $\mathbf{X}^{(i)}$ and $\mathbf{X}^{(j)}$ ($i \neq j$) are orthogonal if

$$\mathbf{X}^{(i)}\mathbf{X}^{(j)^T} = \mathbf{0} , \quad \mathbf{X}^{(i)^T}\mathbf{X}^{(j)} = \mathbf{0} \tag{2.18}$$

In this case, the structure of matrix $\mathbf{X}_V\mathbf{X}_V^T$, the vertical form, see equation (2.14), is simplified as follows:

$$\mathbf{X}_V \mathbf{X}_V^T = \begin{bmatrix} \mathbf{X}^{(1)}\mathbf{X}^{(1)T} & \mathbf{0} & \cdots & \mathbf{0} \\ \mathbf{0} & \mathbf{X}^{(2)}\mathbf{X}^{(2)T} & \cdots & \mathbf{0} \\ \vdots & \vdots & \ddots & \vdots \\ \mathbf{0} & \mathbf{0} & \cdots & \mathbf{X}^{(M)}\mathbf{X}^{(M)T} \end{bmatrix}. \tag{2.19}$$

Thus, in this case, we are dealing with a diagonal matrix and there is no cross-product between trajectory matrices $\mathbf{X}^{(i)}$ and $\mathbf{X}^{(j)}$ in the off-diagonal. Therefore, the total variation of $\mathbf{X}_V$ is distributed among matrices $\{\mathbf{X}^{(1)}, \ldots, \mathbf{X}^{(M)}\}$. Furthermore,

$$r = r_1 + r_2 + \ldots + r_M, \tag{2.20}$$

where, $r_i = \text{rank}\,\mathbf{X}^{(i)}$ $(i = 1, \ldots, M)$ is the trajectory dimension in univariate SSA for each series separately, and $r = \text{rank}\,\mathbf{X}_V$ is the trajectory dimension in the multivariate SSA for all series. It should be mentioned that in SSA terminology r_i is called the number of singular values used in reconstruction stage. In general,

$$r \le r_1 + r_2 + \ldots + r_M. \tag{2.21}$$

Therefore, the equality holds only in the orthogonality situation. Note also that in the case of orthogonality, the structure of matrix $\mathbf{X}_H \mathbf{X}_H^T$ for the HMSSA version is as follows:

$$\mathbf{X}_H \mathbf{X}_H^T = \mathbf{X}^{(1)}\mathbf{X}^{(1)^T} + \cdots + \mathbf{X}^{(M)}\mathbf{X}^{(M)^T}, \tag{2.22}$$

which is exactly similar to those we have for the general case in equation (2.15). Thus, orthogonality does not change the structure of $\mathbf{X}_H \mathbf{X}_H^T$. However, the equality $r = r_1 + r_2 + \ldots + r_M$ holds for HMSSA. Obviously, we do not expect a feedback system if the series are orthogonal. In a feedback system, the multivariate forecasting approach improves the forecasting results of all M series. For example, in a feedback system with two series, $M = 2$, two series are mutually supportive. If the M series are similar according to some criteria, then $r = r_1 = \ldots = r_M$. However, there are several cases where one series is more supportive than the other one. In this case, MSSA provides better results, at least for one of the series, as there are some matched components between series. If all components are matched, then we have similar series, and if there are no matched components, we then have orthogonality between the two. Thus, the following bound for the number of singular values r can be suggested for the VMSSA case:

$$1 \le R_m \le r \le R_m + R_{um} \le ML, \tag{2.23}$$

where, R_m and R_{um} denote the number of matched (at least one) and unmatched components in a whole multivariate system, respectively. Note that for HMSSA, the upper bound is L. The lower bound in equation (2.23) is suitable if the aim is forecasting the series that contains the minimum component, with the use of information from other series, in the whole system. If the aim is forecasting the series with maximum components, then the suitable lower bound is $\text{Max}\{r_i\}_{i=1}^{M}$. In both cases, the lower bounds occur when all series are exactly similar. Furthermore, r hits the upper bound for the case of full orthogonality.

Thus, the ceiling to the multivariate trajectory dimension is simply that obtained when there are, in a sense, no common or matched components among the M series; the presence of matching components reduces the dimension of the multivariate system and indicates that the system is interrelated, which should improve the forecast. The selection of $r < R_m$ leads to a loss of precision as parts of the signals in all series will be lost. On the other hand, if $r > R_m + R_{um}$, then noise is included in the reconstructed series. The selection of $r \cong R_m$ (keeping $r > R_m$) is a good choice for highly interrelated series sharing several common components. The selection of $r \cong R_m + R_{um}$ is necessary when the series being analysed have very little relation to each other.

2.5   Extensions of SSA

2.5.1   SSA-Based Causality Tests

A question that frequently arises in time series analysis is whether one variable can help in predicting another variable. One way to address this question was proposed by Granger in [43] where he formulised the causality concept as follows: process X does not cause process Y if (and only if) the capability to predict the series Y based on the histories of all observable is unaffected by the omission of X's history. The SSA based causality test was first introduced in [44] and is based on the forecasting accuracy and predictability of the direction of change (for more information see [45]).

2.5.1.1   Forecasting Accuracy Criteria

The first criterion is based on out-of-sample forecasting, which is very common in the framework of Granger causality. The main idea is based on the comparison of the forecast values obtained by SSA and MSSA. If the forecasting errors using MSSA are significantly smaller than the forecasting errors of the univariate SSA, it is then concluded that there is a causal relationship between these two series.

Now consider the procedure of constructing a vector of forecasting error for an out-of-sample test in more detail. In the first step, the series $Y_N = (y_1, \ldots, y_N)$ is divided into two separate subseries Y_R and Y_F: $Y_N = (Y_R, Y_F)$, where $Y_R = (y_1, \ldots, y_R)$ and $Y_F = (y_{R+1}, \ldots, y_N)$. The subseries Y_R is used in the reconstruction step to provide the noise-free series $\tilde{Y}_R$. The noise-free series

$\tilde{Y}_R$ is then used for forecasting the subseries Y_F with the help of the recursive formula using SSA and MSSA. The forecast values $\hat{Y}_F = (\hat{y}_{R+1}, \ldots, \hat{y}_N)$ are then used for computing the forecasting errors, and the vector $(y_{R+2}, \ldots, y_N)$ is forecasted using the new subseries $(y_1, \ldots, y_{R+1})$. This procedure is continued recursively up to the end of the series, yielding the h-step-ahead forecasts for univariate and multivariate algorithms. Therefore, the two vectors of h-step-ahead forecasts obtained can be used in examining the association (of order h) between the two series. Let us now consider a formal procedure of constructing a criterion of SSA causality of order h between two arbitrary time series.

Let $X_N = (x_1, \ldots, x_N)$ and $Y_N = (y_1, \ldots, y_N)$ denote two different time series of length N. Set the window lengths L common for both series. Using the embedding terminology, we construct the trajectory matrices $\mathbf{X} = [X_1, \ldots, X_K]$ and $\mathbf{Y} = [Y_1, \ldots, Y_K]$ for the series X_N and Y_N.

Consider an arbitrary loss function $\mathcal{L}$. In econometrics, the loss function $\mathcal{L}$ is usually selected as the mean square error. Let us first assume that the aim is to forecast the series X_N h-step ahead. Thus, the aim is to minimise $\mathcal{L}(X_{K+h} - \hat{X}_{K+h})$, where the vector $\hat{X}_{K+h}$ is an estimate, obtained using a forecasting algorithm, of the vector $X_{K+h} = (x_{K+h}, \ldots, x_{N+h})$. The vector X_{K+h} can be obtained using either univariate SSA or MSSA. Let us first consider the univariate approach.

Define $\Delta_{X_{K+h}} \equiv \mathcal{L}(X_{K+h} - \hat{X}_{K+h})$, where $\hat{X}_{K+h}$ is obtained using univariate SSA; that is, the estimate $\hat{X}_{K+h}$ is obtained only from the vectors $[X_1, \ldots, X_K]$. Let $X_N = (x_1, \ldots, x_N)$ and $Y_{N+d} = (y_1, \ldots, y_{N+d})$ denote two different time series to be considered simultaneously, and consider the same window length L for both series. Here d is an integer, not necessarily non-negative. Now, we forecast $x_{N+1}, \ldots, x_{N+h}$ using the information provided by the series Y_{N+d} and X_N. Next, compute the statistic $\Delta_{X_{K+h}|Y_{K+d}} \equiv \mathcal{L}(X_{K+h} - \tilde{X}_{K+h})$, where $\tilde{X}_{K+h}$ is an estimate of X_{K+h} obtained using MSSA. Now, define the criterion $F_{X|Y}^{(h,d)} = \Delta_{X_{K+h}|Y_{K+d}}/\Delta_{X_{K+h}}$ corresponding to the h step ahead forecast of the series X_N in the presence of the series Y_{N+d}. Note that d is any given integer (even negative). For example, $F_{X|Y}^{(h,0)}$ means that $d = 0$ and that we use the series X_N and Y_N simultaneously (with zero lag).

If $F_{X|Y}^{(h,d)}$ is small, then having information obtained from the series Y helps us to have better forecasts of the series X. This means there is a relationship between series X and Y of order h according to this criterion, and the predictions using multivariate SSA are much more accurate than the predictions by univariate SSA. If $F_{X|Y}^{(h,d)} < 1$, then we conclude that the information provided by the series Y can be regarded as useful or *supportive* for forecasting the series X. Alternatively, if the values of $F_{X|Y}^{(h,d)} \geq 1$, then, either there is no detectable association between X and Y or the performance of univariate

SSA is better than MSSA (this may happen, for example, when the series Y has structural breaks misdirecting the forecasts of X).

To asses which series is more supportive in forecasting, we need to consider another criterion. We obtain $F_{Y|X}^{(h,d)}$ in a similar manner. Now, these measures tell us whether using extra information about time series Y_{N+d} (or X_{N+d}) supports X_N (or Y_N) in h-step forecasting. If $F_{Y|X}^{(h,d)} < F_{X|Y}^{(h,d)}$, we then conclude that X is more supportive to Y than Y to X. Otherwise, if $F_{X|Y}^{(h,d)} < F_{Y|X}^{(h,d)}$, we conclude that Y is more supportive to X than X to Y.

However, if $F_{Y|X}^{(h,d)} < 1$ and $F_{X|Y}^{(h,d)} < 1$, it is then concluded that X and Y are *mutually supportive*. This is called F-feedback (forecasting feedback) which means that the use of a multivariate system improves the forecasting for both series.

2.5.1.2   Direction of Change Based Criteria

The direction of change criterion shows the proportion of forecasts that correctly predict the direction of the series movement. For the forecasts obtained using only X_N (univariate case), let Z_{X_i} take the value 1 if the forecast correctly predicts the direction of change, and 0 otherwise. Then, $\bar{Z}_X = \sum_{i=1}^{n} Z_{Xi}/n$ shows the proportion of forecasts that correctly predict the direction of the series. The Moivre–Laplace central limit theorem [45] implies that, for large samples, the test statistic $2(\bar{Z}_X - 0.5)N^{1/2}$ is approximately distributed as standard normal. When $\bar{Z}_X$ is significantly larger than 0.5, then the method is said to have the ability to predict the direction of change. Alternatively, if $\bar{Z}_X$ is significantly smaller than 0.5, the method tends to give the wrong direction of change.

For the multivariate case, let $Z_{X|Y}$ take a value 1 if the forecast of the direction of change of the series X having information about the series Y is correct, and 0 otherwise. Then, we define the following criterion: $D_{X|Y}^{(h,d)} = \bar{Z}_X/\bar{Z}_{X|Y}$, where h and d have the same interpretation as for $F_{X|Y}^{(h,d)}$. The criterion $D_{X|Y}^{(h,d)}$ characterises the improvement we achieve from the information contained in Y_{N+h} (or X_{N+h}) for forecasting the direction of change in the h step ahead forecast.

If $D_{X|Y}^{(h,d)} < 1$, then, having information about series Y helps us to have better prediction of the direction of change for series X. Alternatively, if $D_{X|Y}^{(h,d)} > 1$, then, univariate SSA performs better than the multivariate version. Similar to forecasting accuracy criterion, we can define a feedback system and check which series is more supportive based on the predictability of the direction of change.

To test the significance of the value of $D_{X|Y}^{(h,d)}$, one can use the test developed in [45]. As in the comparison of two proportions, when the hypothesis about the difference between two proportions is tested, here it is first tested whether the two proportions are dependent. The test is different depending on whether the proportions are independent or dependent. In our case, obviously, Z_X and $Z_{X|Y}$ are dependent. Let us consider the test statistic for the difference between the forecast values of Z_X and $Z_{X|Y}$ as arranged in Table 2.2. The test

TABLE 2.2: An arrangement of Z_X and $Z_{X|Y}$ in forecasting n future points of the series X.

$Z_{X|Y}$	Z_X	number
1	1	a
1	0	b
0	1	c
0	0	d
Total		$n = a + b + c + d$

is valid when the average of the discordant cell frequencies, $(b+c)/2$, is equal or more than 5. However, when it is less than 5, a binomial test can be used. The estimated proportion of the correct forecasts using the multivariate system is $P_{X|Y} = (a + b)/n$, and the estimated proportion using the univariate version is $P_X = (a + c)/n$. The difference between the two estimated proportions is

$$\pi = P_{X|Y} - P_X = \frac{a+b}{n} - \frac{a+c}{n} = \frac{b-c}{n} \tag{2.24}$$

Since the two population probabilities are dependent, the estimated standard error is defined as follows:

$$S\hat{E}(\pi) = \frac{1}{n}\sqrt{(b+c) - \frac{(b-c)^2}{n}}. \tag{2.25}$$

Writing then the null and alternative hypotheses as:

$$\begin{aligned} H_0 &: \pi_d = \Delta_0 \\ H_a &: \pi_d \neq \Delta_0 \end{aligned} \tag{2.26}$$

The test statistic, assuming that the sample size is large enough for having the Normal approximation to the Binomial distribution, is:

$$T_{\pi_d} = \frac{\pi - \Delta_0 - 1/n}{S\hat{E}(\pi)} \tag{2.27}$$

where $1/n$ is the continuity correction. In our case $\Delta_0 = 0$, and the test statistic becomes:

$$T_{\pi_d} = \frac{(b-c)/n - 1/n}{1/n\sqrt{(b+c) - (b-c)^2/n}} = \frac{b-c-1}{\sqrt{(b+c) - (b-c)^2/n}} \tag{2.28}$$

Under the null hypothesis of no difference between samples Z_X and $Z_{X|Y}$, T_{π_d} has standard normal distribution.

2.5.2   Change Point Detection

The SSA technique and consequently the change-point detection algorithm are nonparametric techniques. However, under certain conditions, the SSA change point detection algorithm can be considered as a proper statistical procedure [1]. One of the features of the SSA algorithm is that the distance between the vectors X_j $(j = 1, \ldots, K)$ and the r-dimensional hyperplane is controlled by the choice of r and can be reduced to a rather small value. If the time series $Y_N = (y_1, \ldots, y_N)$ is continued for $h > N$ and there is no change in the mechanism which generates the values y_t, then this distance should stay reasonably small for X_j $j \geqslant K$. The Q such vectors are evaluated for testing this hypothesis. However, if at a certain time $N + h$ the mechanism generating y_t $t \geqslant N + h$ has changed, then an increase in the distance between the r-dimensional hyperplane and the vectors X_j $j \geqslant K + h$ is to be expected. If there are no structural changes, then the SSA continuation of the time series should agree with the continued series. That is, the Q vectors X_j $j \geqslant K$ should stay close to the r-dimensional subspace. A change in structure of the time series should force the corresponding vectors X_j out of the subspace. This is the central idea to the change point detection algorithm proposed in [1].

2.5.3   Automated Selection of L and r

The aim of the automated selection of L and r is to enable users who are not overly conversant with the SSA technique to still be able to exploit the method for their filtering and forecasting problems. Particularly, the forecasting algorithm is optimised based on a loss function such that it provides users with the best SSA choices for obtaining forecasts with a minimum error. The automated SSA forecasting algorithm is presented below, and in doing so we mainly follow [6].

1. Consider a real-valued nonzero time series $Y_N = (y_1, \ldots, y_N)$ of length N and divide it into two parts; $\frac{2}{3}^{rd}$ of observations for model training and testing, and $\frac{1}{3}^{rd}$ for validating the selected model.

2. Use the training data to construct the trajectory matrix $\mathbf{X}$. Initially, we begin with $L = 2$ and in the process, evaluate all possible values of L $(2 \leq L \leq \frac{N}{2})$.

3. Obtain the SVD of $\mathbf{X}$ and evaluate all possible combinations of r $(1 \leq r \leq L - 1)$ singular values (step by step) for the selected L and split the elementary matrices $\mathbf{X}_i$ $(i = 1, \ldots, L)$ into several groups and sum the matrices within each group.

4. Perform diagonal averaging to transform the matrix with the selected r singular values into a Hankel matrix which can then be converted into

a time series. The steps up to this stage filter the noisy series and the output, which is a noise reduced series that can be used for forecasting new data points.

5. Define a loss function $\mathcal{L}$ proportional to the distance between the desired and estimated vectors.

6. When forecasting a series Y_N h-step ahead, the forecast error is minimised by setting $\mathcal{L}(X_{K+h} - \hat{X}_{K+h})$ where the vector $\hat{X}_{K+h}$ contains the h-step ahead of forecasts obtained using the SSA forecasting algorithm.

7. Find the combination of L and r which minimises $\mathcal{L}$ and thus represents the optimal SSA choices (L and r).

8. Finally, use the optimal L to decompose the series comprised of the validation set, and then select r singular values for reconstructing the less noisy time series, and use this newly reconstructed series for forecasting new data points.

2.5.4   SSA Based on Minimum Variance

The SSA technique based on a minimum variance estimator has been introduced in [33]. Basic SSA considers the least squares (LS) estimator for filtering and forecasting noisy time series. The minimum variance (MV) estimator was introduced as an alternative. As noted in [33], the minimum variance estimator is recognized as the optimal linear estimator which provides minimum total residual power. Presented below is the reconstruction with SSA based on the minimum variance estimator, and in doing so we mainly follow [34].

Consider a noisy time series Y_N of length N. Then, the signal-plus-noise series without any assumptions about the nature of the signal will be:

$$Y_N = S_N + E_N, \tag{2.29}$$

where S_N represents the signal component and E_N the noise component. The corresponding trajectory matrices can be written as:

$$\mathbf{X}_y = \mathbf{X}_s + \mathbf{X}_e. \tag{2.30}$$

Then the SVD of the matrix $\mathbf{X}_y$ can be written as:

$$\mathbf{X}_y = \mathbf{U}_y \mathbf{\Sigma}_y \mathbf{V}_y^T = \begin{bmatrix} \mathbf{U}_s & \mathbf{U}_e \end{bmatrix} \begin{bmatrix} \mathbf{\Sigma}_s & 0 \\ 0 & \mathbf{\Sigma}_e \end{bmatrix} \begin{bmatrix} \mathbf{V}_s^T \\ \mathbf{V}_e^T \end{bmatrix}, \tag{2.31}$$

where $\mathbf{U}_s \in \mathbb{R}^{L \times r}$, $\mathbf{\Sigma}_s \in \mathbb{R}^{r \times r}$ and $\mathbf{V}_s \in \mathbb{R}^{K \times r}$. The SVD of the Hankel matrix of the signal $\mathbf{X}_s$ can be represented as:

$$\mathbf{X}_s = \begin{bmatrix} \mathbf{U}_s & \mathbf{U}_e \end{bmatrix} \begin{bmatrix} \mathbf{U}_s & 0 \\ 0 & 0 \end{bmatrix} \begin{bmatrix} \mathbf{V}_s^T \\ \mathbf{V}_e^T \end{bmatrix}. \tag{2.32}$$

It is clear that the Hankel matrix $\mathbf{X}_s$ cannot be reconstructed exactly if it is perturbed by noise. To remove the effect of the noise term, it is assumed that the vector space of the noisy time series can be split in mutually orthogonal noise and signal + noise subspaces. The components in the noise subspace are suppressed or even removed completely. Therefore, one can reconstruct the noise free series from signal+noise subspace by choosing the weight. Thus, by adapting the weights of the different singular components, an estimate of the Hankel matrix $\mathbf{X}_s$, which corresponds to a noise-reduced series, can be achieved:

$$\mathbf{X}_s = \mathbf{U}(\mathbf{W}\boldsymbol{\Sigma})\mathbf{V}^T, \tag{2.33}$$

where $\mathbf{W}$ is the diagonal matrix containing the weights. Now, the problem is choosing the weight matrix $\mathbf{W}$. Consider the weight matrix $\mathbf{W}$ based on the LS and MV estimates. The LS and MV estimates can be defined based on the weight matrix $\mathbf{W}_{r \times r}$ as follows:

$$\widehat{\mathbf{X}}_{sLS} = \mathbf{U}_s(\mathbf{W}_{LS}\boldsymbol{\Sigma}_s)\mathbf{V}_s^T \tag{2.34}$$

$$\widehat{\mathbf{X}}_{sMV} = \mathbf{U}_s(\mathbf{W}_{MV}\boldsymbol{\Sigma}_s)\mathbf{V}_s^T \tag{2.35}$$

where $\mathbf{W}_{LS} = \mathbf{I}_{r \times r}$

$$\mathbf{W}_{MV} = diag\left(\left(1 - \frac{\sigma_{noise}^2}{\lambda_1^2}\right), \ldots, \left(1 - \frac{\sigma_{noise}^2}{\lambda_r^2}\right)\right) \tag{2.36}$$

2.5.5 SSA Based on Perturbation Theory

SSA based on perturbation theory was developed following the introduction of SSA based on the minimum variance estimate. The rationale was to provide a comparatively better reconstruction by removing the noise term from the matrices attainable via both SSA based on least squares and SSA based on minimum variance. SSA based on perturbation theory is concisely presented below by following [34]. Let us begin by defining three matrices as

$$\tilde{\mathbf{U}}_s = \mathbf{U}_s + \delta\mathbf{U}_s \tag{2.37}$$

$$\tilde{\mathbf{V}}_s = \mathbf{V}_s + \delta\mathbf{V}_s \tag{2.38}$$

$$\tilde{\boldsymbol{\Sigma}}_s = \boldsymbol{\Sigma}_s + \delta\boldsymbol{\Sigma}_s. \tag{2.39}$$

The ideal scenario is where we can remove the noise terms $\delta\mathbf{U}_s$, $\delta\mathbf{V}_s$ and $\delta\boldsymbol{\Sigma}_s$. It is noteworthy that in SSA based on the minimum variance estimator we seek to remove $\delta\boldsymbol{\Sigma}_s$ whilst in basic SSA we keep all these noisy terms in the SVD expansion. In general, an estimation of the signal matrix $\mathbf{X}_s$ can be presented as:

$$\widehat{\mathbf{X}}_s = (\mathbf{G}_U\tilde{\mathbf{U}}_s)(\mathbf{G}_\Sigma\tilde{\boldsymbol{\Sigma}}_s)(\mathbf{G}_V\tilde{\mathbf{V}}_s^T) \tag{2.40}$$

where matrices $\mathbf{G}_U$, $\mathbf{G}_V$ and $\mathbf{G}_\Sigma$ based on perturbation theory are presented in Table 2.3 where $\mathbf{I}$ is an identity matrix and matrices $\mathbf{P}_U$, $\mathbf{P}_\Sigma$ and $\mathbf{P}_V$ are obtained up to second order perturbation theory.

TABLE 2.3: Matrices $\mathbf{G}_U$, $\mathbf{G}_\Sigma$ and $\mathbf{G}_V$ for different estimator.

Estimation Method	$\mathbf{G}_U$	$\mathbf{G}_\Sigma$	$\mathbf{G}_V$
Least Square	$\mathbf{I}$	$\mathbf{I}$	$\mathbf{I}$
Minimum Variance	$\mathbf{I}$	$\mathbf{W}_{MV}$	$\mathbf{I}$
Perturbation Theory	$\mathbf{P}_U$	$\mathbf{P}_\Sigma$	$\mathbf{P}_V$

2.5.6 Two Dimensional SSA

Two Dimensional SSA (2D-SSA) is an extension of MSSA [35]. It has previously been used for processing two-dimensional scalar fields [1], image processing [36] and time series analysis [35]. The advantage of 2D-SSA for biomedical science stems from its ability of analysing data from multiple objects simultaneously. In what follows the 2D-SSA approach is briefly explained, and in doing so we mainly follow [35].

Consider a two dimensional dataset, $F:\{y_{11}, ..., y_{MN}\} \times \{X_1, ..., X_t\}$, where X_t can for example be some form of indicator, M indicates the number of data sets relating to different objects of interest and N denotes the number of different objects. The other stages the 2D-SSA procedure are similar to univariate SSA and as such are not reproduced here. However, the key difference between 2D-SSA and SSA appears in relation to selecting the SSA choice of L. This is because in 2D-SSA analysis two different values for the window length L (L_1, L_2) are considered, whilst in univariate SSA only a single window length is required. In that context, if $L_2 = 1$, then 2D-SSA is equivalent to SSA. If $L_2 = M$, the interaction among different objects of interest will be taken into consideration and one object composite index will reflect the other objects, fluctuation and information. At the same time, if $L_2 = M \times N$, then 2D-SSA is equivalent to MSSA.

2.5.7 SSA Based on L_1-Norm

The basic version of SSA which has been explained thus far is based on the *Frobenius* norm. In fact, in [37] it was proven that SSA based on the L_2-norm is sensitive to outliers via varying impacts on reconstruction and forecasting. In simple terms, this is because the L_2-norm is concerned with seeking an average or mean which is by definition affected by outliers. However, the reconstruction stage of SSA can be reformed based on the L_1-norm. In this case, one should seek to find and exploit the median, which in turn should result in little or no sensitiveness to outliers.

Let $\mathcal{H}\mathbf{A}$ be the result of the Hankelisation of matrix $\mathbf{A}$. The Hankel matrix

$\mathcal{H}\mathbf{A}$ uniquely defines the original series by relating the values in the anti-diagonals to the values in the series. In basic SSA which is based on L_2-norm, the Hankelisation procedure is obtained via diagonal averaging whilst L_1-Hankelisation corresponds to compute median of the matrix elements over the "antidiagonal" using the following L_1-Hankelisation definition. Let $\mathbf{A}$ be a $L \times K$ matrix and $s = i + j$　$(2 \leq s \leq L + K)$, then the element $\tilde{a}_{ij}$ of the matrix $\mathcal{H}\mathbf{A}$ with respect to L_1-norm is

$$\tilde{a}_{ij} = \underset{(l,k)\in A_s}{\mathrm{median}}\, a_{lk}, \qquad\qquad (2.41)$$

where $A_s = \{(l, k) : l + k = s, 1 \leq l \leq L, 1 \leq k \leq K\}$.

In the corresponding chapters of this book we will see how statistical and physical constraints can be exploited to develop an adaptive technique for the grouping process of SSA. This concept is nicely used in developing a new adaptive line enhancer to separate the cyclic information from the noisy observations. In another front we will see how a tensor can be built from single or multichannel data and a multiway (tensor decomposition) approach is proposed for the data decomposition instead of conventional SVD. On the other hand, the idea of an augmented covariance matrix decomposition will be discussed to exploit the correlation between the real and imaginary parts of complex data in developing a robust SSA for the decomposition of complex signals.

Bibliography

[1] Golyandina, N., Nekrutkin, V., and Zhigljavsky, A. (2001). *Analysis of Time Series Structure: SSA and related techniques.* CRC Press.

[2] Thomakos, D., Wang, T., and Wille, L. (2002). Modeling Daily Realized Futures Volatility Using Singular Spectrum Analysis. *Physica A: Statistical Mechanics and its Application,* **312**(3), 505–519.

[3] Rodríguez-Aragón, L. J., and Zhigljavsky, A. (2010). Singular Spectrum Analysis for Image Processing. *Statistics and Its Interface,* **3**(3), 419–426.

[4] Hassani, H., Heravi, S., and Zhigljavsky, A. (2009). Forecasting European Industrial Production with Singular Spectrum Analysis. *International Journal of Forecasting,* **25**(1), 103–118.

[5] Cañóna, J., Domíngueza, F., and Valdés, J. B. (2011). Downscaling Climate Variability Associated with Quasi-periodic Climate Signals: A New Statistical Approach Using MSSA. *Journal of Hydrology,* **398**(1C2), 65–75.

[6] Hassani, H., Webster, A., Silva, E. S., and Heravi, S. (2015). Forecasting U.S. tourist arrivals using optimal Singular Spectrum Analysis. *Tourism Management,* **46**, 322–335.

[7] Broomhead, D. S., and King, G. P. (1986). On the qualitative analysis of experimental dynamical systems, In *Nonlinear Phenomena and Chaos,* 113–144. Adam Hilger, Bristol, England.

[8] Telesca, L., Matcharasvili, T., Chelidze, T., and Zhukova, N. (2012). Relationship between seismicity and water level in the Enguri high dam area (Georgia) using the singular spectrum analysis. *Natural Hazards and Earth System Science,* **12**(8), 2479–2485.

[9] Sanei, S., and Lee, T. K. (2011). Detection of periodic signals using a new adaptive line enhancer based on singular spectrum analysis. In *Information, Communications and Signal Processing (ICICS), 8th International Conference,* 1–5.

[10] Hou, Z., Wen, G., Tang, P., and Cheng, G. (2014). Periodicity of Carbon Element Distribution Along Casting Direction in Continuous-Casting

Billet by Using Singular Spectrum Analysis. *Metallurgical and Materials Transactions B*, **45**(5), 1817–1826.

[11] Sanei, S., Ghodsi, M., and Hassani, H. (2011). An adaptive singular spectrum analysis approach to murmur detection from heart sounds. *Medical Engineering and Physics*, **33**(3), 362–367.

[12] Liu, K., Law, S. S., Xia, Y., and Zhu, X. Q. (2014). Singular spectrum analysis for enhancing the sensitivity in structural damage detection. *Journal of Sound and Vibration*, **333**(2), 392–417.

[13] Mirabedini, A. S., Karrabi, M., and Ghasemi, I. (2013). Viscoelastic behaviour of NBR/phenolic compounds. *Iranian Polymer Journal*, **22**(1), 25–32.

[14] Chen, Q., Dam, T. V., Sneeuw, N., Collilieux, X., Weigelt, M., and Rebischung, P. (2013). Singular spectrum analysis for modelling seasonal signals from GPS time series. *Journal of Geodynamics*, **72**, 25–35.

[15] Aydin, S., Saraoglu, H. M., and Kara, S. (2011). Singular Spectrum Analysis of Sleep EEG in Insomnia. *Journal of Medical Systems*, **35**(4), 457–461.

[16] Ghodsi, M., Hassani, H., and Sanei, S. (2010). Extracting Fetal Heart Signal from Noisy Maternal ECG by Singular Spectrum Analysis. *Statistics and Its Interface*, **3**(3), 399–411.

[17] Patterson, K., Hassani, H., Heravi, S., and Zhigljavsky, A. (2011). Forecasting the final vintage of the industrial production series. *Journal of Applied Statistics*, **38**(10), 2183–2211.

[18] Vautard, R., and Ghill, M. (1989). Singular spectrum analysis in nonlinear dynamics with applications to paleoclimatic time series. *Physica D: Nonlinear Phenomena*, **35**(3), 395–424.

[19] Hassani, H., Soofi, A., and Zhigljavsky, A. (2013). Predicting Inflation Dynamics with Singular Spectrum Analysis. *Journal of the Royal Statistical Society Series*, **176**(3), 743–760.

[20] Zhang, Y., and Hui, X. F. (2012). Research on daily exchange rate forecasting with multivariate singular spectrum analysis. In *Management Science and Engineering (ICMSE), 2012 International Conference on*, 1365–1370.

[21] Mohammad, Y., and Nishida, T. (2011). Discovering causal change relationships between processes in complex systems. In *System Integration (SII), 2011 IEEE/SICE International Symposium on*, 12–17.

[22] Zhao, X., Shang, P., and Qiuyue, J. (2011). Multifractal detrended cross-correlation analysis of Chinese stock markets based on time delay. *Fractals*, **19**(3), 329–338.

[23] Kapl, M., and Mueller, W. (2010). Prediction of steel prices: a comparison between a conventional regression model and MSSA. *Stat Interface*, **3**(3), 369–375.

[24] Oropeza, V., and Sacchi, M. (2011). Simultaneous seismic data denoising and reconstruction via multichannel singular spectrum analysis. *Geophysics*, **76**(3), 25–32.

[25] Groth, A., and Ghil, M. (2011). Multivariate singular spectrum analysis and the road to phase synchronization. *Physical Review E*, **84**(3), 036206.

[26] Cana, J., Domngueza, F., and Valds, J. B. (2011). Downscaling Climate Variability Associated with Quasi-periodic Climate Signals: A New Statistical Approach Using MSSA. *Journal of Hydrology*, **398**(12), 65–75.

[27] Hassani, H., Heravi, S., and Zhigljavsky, A. (2013). Forecasting UK Industrial Production with Multivariate Singular Spectrum Analysis. *Journal of Forecasting*, **32**(5), 395–408.

[28] Mahmoudvand, R., and Zokaei, M. (2012). On the singular values of the Hankel matrix with application in singular spectrum analysis. *Chilean Journal of Statistics*, **3**(1), 43–56.

[29] Stepanov, D., and Golyandina, N. (2005). SSA-based approaches to analysis and forecast of multidimensional time series. *In: Proceedings of the 5th St.Petersburg Workshop on Simulation, June 26-July 2, 2005*, St. Petersburg State University, St. Petersburg, 293–298.

[30] Rao, C. R., and Mitra, S. K. (1971). *Generalized Inverse of Matrices and Its Applications*. Wiley, NewYork.

[31] Hassani, H., Mahmoudvand, R., and Zokaei, M. (2011). Separability and Window Length in Singular Spectrum Analysis. *Comptes Rendus Mathematique*, **349**(17-18), 987–990.

[32] Mahmoudvand, R., Najari, N., and Zokaei, M. (2013). On the Optimal Parameters for Reconstruction and Forecasting in Singular Spectrum Analysis. *Communication in Statistics: Simulations and Computations*, **42**(4), 860–870.

[33] Hassani, H. (2010). Singular spectrum analysis based on the minimum variance estimator. *Nonlinear Analysis: Real World Applications*, **11**(3), 2065–2077.

[34] Hassani, H., and Thomakos, D. (2010). A review on singular spectrum analysis for economic and financial time series. *Statistics and Its Interface*, **3**, 377–397.

[35] Zhang, J., Hassani, H., Xie, H., and Zhang, X. (2014). Estimating multi-country prosperity index: A two-dimensional singular spectrum analysis approach. *Journal of Systems Science and Complexity*, **27**(1), 56–74.

[36] Golyandina, N., and Usevich, K. (2010). 2D-extension of Singular Spectrum 151 Analysis: algorithm and elements of theory. *Matrix Methods: Theory, Algorithms, Applications*, World Scientific Publishing, 450-474.

[37] Hassani, H., Mahmoudvand, R., Omer, H. N., and Silva, E. S. (2014). A Preliminary Investigation into the Effect of Outlier(s) on Singular Spectrum Analysis. *Fluctuation and Noise Letters*, **13**(4), 1450029(123).

[38] Sanei, S., Lee, T. K. M., and Abolghasemi, V. (2012). A New Adaptive Line Enhancer Based on Singular Spectrum Analysis. *IEEE Transactions on Biomedical Engineering*, **59**(2), 428–433.

[39] Kouchaki, S., Sanei, S., Arbon, E. L., and Dijk, D. J. (2015). Tensor Based Singular Spectrum Analysis for Automatic Scoring of Sleep EEG. *IEEE Transactions on Neural Systems and Rehabilitation Engineering*, **23**(1), 1–9.

[40] Corsini, J., Shoker, L., Sanei, S., and Alarcon, G. (2006). Epileptic seizure prediction from scalp EEG incorporating BSS. *Biomedical Engineering, IEEE Transactions*, **53**(5), 790–799.

[41] Harshman, R. A. (1970). Foundations of the PARAFAC procedure: Model and conditions for an explanatory multimode factor analysis. *UCLA Working Papers in Phonetics*, **16**, 1–84.

[42] Hassani, H., and Mahmoudvand, R. (2013). Multivariate Singular Spectrum Analysis: A General View and New Vector Forecasting Approach. *International Journal of Energy and Statistics*, **1**(1), 55–83.

[43] Granger, C. W. J. (1980). Testing for causality: A personal viewpoint. *Journal of Economic Dynamics and Control*, **2**, 329-352.

[44] Hassani, H., Zhigljavsky, A., Patterson, K., Soofi, A. (2010). A Comprehensive Causality Test Based on the Singular Spectrum Analysis. *Causality in Science*, 379–406.

[45] Hassani, H., Heravi, S., and Zhigljavsky, A. (2013). Forecasting UK Industrial Production with Multivariate Singular Spectrum Analysis. *Journal of Forecasting*, **32**(5), 395–408.

Chapter 3

SSA Application to Sleep Scoring

3.1 Introduction

Sleep is a physiological and biological state of almost all inhabitants on the earth [1]. Sleep has been described by early scientists as a passive condition where the brain is isolated from the rest of the body. Alcmaeon claimed that sleep is caused by the blood receding from the blood vessels in the skin to the interior parts of the body. Philosopher and scientist Aristotle suggested that while food is being digested, vapors rise from the stomach and penetrate into the head. As the brain cools, the vapors condense, flow downward and then cool the heart, which causes sleep. Some others still claim that toxins that poison the brain cause sleep [2]. With the discovery of brain waves and later the discovery of the EEG system, the way sleep was studied changed.

Sleep is the state of natural rest observed in humans and animals, and even invertebrates such as the fruit fly *Drosophila*. Lack of sleep seriously influences our brain's ability to function. With continued lack of sufficient sleep, the part of the brain that controls language, memory, planning and sense of time is severely affected and the ability of judgement deteriorates. Sleep is an interesting and not perfectly known physiological phenomenon. Sleep state has become important evidence for diagnosing mental disease and psychological abnormality. Sleep is characterized by a reduction in voluntary body movement, decreased reaction to external stimuli, an increased rate of anabolism (the synthesis of cell structures), and a decreased rate of catabolism (the breakdown of cell structures). Sleep is necessary for the life of most creatures. The capability for arousal from sleep is a protective mechanism and also necessary for health and survival. In terms of physiological changes in the body, and particularly changes in the state of the brain, sleep is different from unconsciousness [3]. However, in manifestation, sleep is defined as a state of unconsciousness from which a person can be aroused. In this state, the brain is relatively more responsive to internal stimuli than external stimuli. Sleep should be distinguished from coma. Coma is an unconscious state from which a person cannot be aroused.

Historically, sleep was thought to be a passive state. However, sleep is now known to be a dynamic process, and our brains are active during sleep. Sleep affects our physical and mental health, and the immune system.

States of physiological and metabolic activity of the human body, particularly the brain, during sleep and wakefulness resulted from different activating and inhibiting forces that are generated within the brain and the changes in blood biochemistry. These mutually influence other states of body activity such as breathing and heart rate.

Neurotransmitters (chemicals involved in nerve signalling) control whether one is asleep or awake by acting on nerve cells (neurons) in different parts of the brain. Neurons located in the brainstem actively cause sleep by inhibiting other parts of the brain that keep a person awake.

In human beings, it has been demonstrated that the metabolic activity of the brain decreases significantly after 24 hours of sustained wakefulness. Sleep deprivation results in a decrease in body temperature, a decrease in immune system function as measured by white blood cell count (the soldiers of the body), and a decrease in the release of growth hormone. Sleep deprivation can also cause increased heart rate variability [4].

Sleep is necessary for the brain to remain healthy. Sleep deprivation makes a person drowsy and unable to concentrate. It also leads to impairment of memory and physical performance, and a reduced ability to carry out mathematical calculations and other mental tasks. If sleep deprivation continues, hallucinations and mood swings may develop. Sleep deprivation not only has a major impact on cognitive functioning, but also on emotional and physical health. Disorders such as sleep apnea which result in excessive daytime sleepiness have been linked to stress and high blood pressure. Research has also suggested that sleep loss may increase the risk of obesity because chemicals and hormones that play a key role in controlling appetite and weight gain are released during sleep.

Release of growth hormone in children and young adults takes place during deep sleep. Most cells of the body show increased production and reduced breakdown of proteins during deep sleep. Sleep helps humans maintain optimal emotional and social functioning while we are awake by giving rest during sleep to the parts of the brain that control emotions and social interactions.

3.2 Stages of Sleep

Sleep is a dynamic process. Loomis provided the earliest detailed description of various stages of sleep in the mid-1930s, and in the early 1950s Aserinsky and Kleitman identified rapid eye movement (REM) sleep [3]. There are two distinct states that alternate in cycles and reflect differing levels of neuronal activity. Each state is characterized by a different type of EEG activity. Sleep consists of nonrapid eye movement (NREM) and REM sleep. NREM is further subdivided into four stages of I (Drowsiness), II (Light Sleep), III (Deep Sleep) and IV (Very Deep Sleep).

During the night the NREM and REM stages of sleep alternate. Stages I, II, III, and IV are followed by REM sleep. A complete sleep cycle, from the beginning of stage I to the end of REM sleep, usually takes about one and a half hours. However, generally, the ensuing sleep is relatively short and, in practice, often 10–30 minutes duration suffices.

3.2.1 NREM Sleep

The first stage, stage I, is the stage of drowsiness and very light sleep, which is considered as a transition between wakefulness and sleep. During this stage, the muscles begin to relax. It occurs upon falling asleep and during brief arousal periods within sleep, and usually accounts for 5–10% of total sleep time. An individual can be easily awakened during this stage. Drowsiness shows marked age-determined changes. Hypnagogic rhythmical 4–6/sec theta activity of late infancy and early childhood is a significant characteristic of such ages. Later in childhood, and in several cases, in the declining years of life, the drowsiness onset involves larger amounts of slow activity mixed with the posterior alpha rhythm [5]. In adults however, the onset of drowsiness is characterised by gradual or brisk alpha dropout [5]. The slow activity increases as the drowsiness becomes deeper. Other findings show that in light drowsiness the P300 response increases in latency and decreases in amplitude [6], and the inter- and intra-hemispheric EEG coherence alter [7]. Figure 3.1 shows a set of EEG signals recorded during the state of drowsiness. The seizure type activity within the signal is very clear.

Deep drowsiness involves appearance of vertex waves. Before the appearance of the first spindle trains, vertex waves occur (transition from stage I to II). These sharp waves are also known as parietal humps [8]. The vertex wave is a compound potential; a small spike discharge of positive polarity followed by a large negative wave, which is a typical discharge wave. It may occur as an isolated event with larger amplitude than that of normal EEG. In aged individuals they may become small, inconspicuous and hardly visible. Another signal feature for deep drowsiness is the positive occipital sharp transients of sleep (POST).

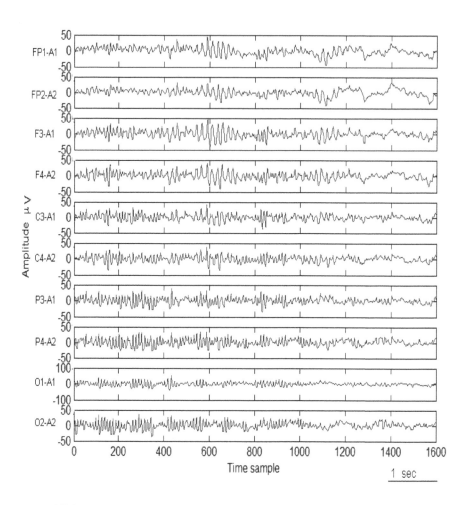

FIGURE 3.1: Examplar EEG signals recorded during drowsiness.

Spindles (also called sigma activities), are the trains of barbiturate-induced beta activity, which occur independently in approximately 18–25 cycles/sec predominantly in the frontal lobe of the brain. They may be identified as a group of rhythmic waves characterised by progressively increasing, then gradually decreasing amplitude [5]. However, the use of middle electrodes shows a very definite maximum of the spindles over the vertex during the early stages of sleep.

Stage II of sleep occurs throughout the sleep period and represents 40–50% of the total sleep time. During Stage II, the brain waves slow down, with occasional bursts of rapid waves. Eye movement stops during this stage. Slow frequencies ranging from 0.7 to 4 cycles/sec are usually predominant; their voltage is high with a very prominent occipital peak in small children that gradually falls when they become older.

K-complexes appear in Stage II and constitute a significant response to arousing stimuli. As to the topographical distribution over the brain, the K-complex shows a maximum over the vertex and has presence around the frontal midline [5]. As to the wave morphology, the K- complex consists of an initial sharp component, followed by a slow component that fuses with a superimposed fast component.

In Stage III, delta waves begin to appear. They are interspersed with smaller, faster waves. Sleep spindles are still present in approximately 12–14 cycles/sec but gradually disappear as the sleep becomes deeper. Figure 3.2 illustrates segments of sleep EEG signals from four electrodes of EEG.

In Stage IV, delta waves are the primary waves recorded from the brain. Delta or slow wave sleep (SWS) usually is not seen during a routine EEG [9]. However, it is seen during prolonged (> 24 hours) EEG monitoring.

Stages III and IV are often distinguished from each other only by the percentage of delta activity. Together, they represent up to 20% of total sleep time. During Stages III and IV all eye and muscle movement ceases. It is difficult to wake up someone during these two stages. If someone is awakened during deep sleep, he does not adjust immediately and often feels groggy and disoriented for several minutes after waking up. Generally, analysis of EEG morphology during stage IV has been of less interest since the brain functionality cannot be examined during this period easily

3.2.2 REM Sleep

REM sleep including 20–25% of the total sleep follows NREM sleep and occurs 4–5 times during a normal 8 to 9-hour sleep period. The first REM period of the night may be less than 10 minutes in duration, while the last period may exceed 60 minutes.

In extremely sleepy individuals, the duration of each bout of REM sleep is very short or it may even be absent. REM sleep is usually associated with dreaming. During REM sleep, the eyeballs move rapidly, the heart rate and breathing become rapid and irregular, the blood pressure rises, and there is loss

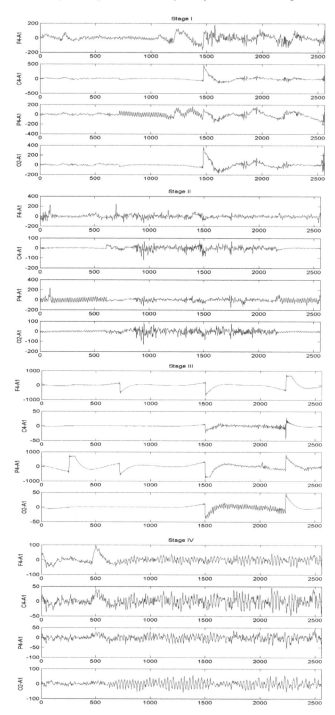

FIGURE 3.2: Segments of sleep EEG signals from four electrodes of EEG;
(a) Stage I, (b) Stage II, (c) Stage III, and (d) Stage IV.

of muscle tone (paralysis), i.e. the muscles of the body are virtually paralysed. The brain is highly active during REM sleep, and the overall brain metabolism may be increased by as much as 20%. The EEG activity recorded in the brain during REM sleep is similar to that recorded during wakefulness. In a patient with REM sleep behaviour disorder (RBD), the paralysis is incomplete or absent, allowing the person to act out his dreams, which can be vivid, intense, and violent. These dream-acting behaviours include talking, yelling, punching, kicking, sitting, jumping from the bed, arm flailing and grabbing. Although the RBD may occur in association with different degenerative neurological conditions the main cause is still unknown.

Evaluation of REM sleep involves a long waiting period since the first phase of REM does not appear before 60–90 minutes after the start of sleep. The EEG in the REM stage shows low voltage activity with slower rate of alpha. Figure 3.3 shows the brain waves during REM sleep.

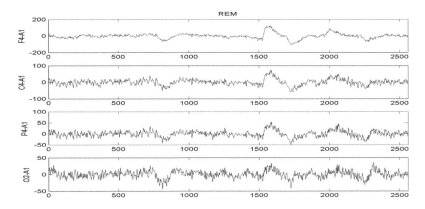

FIGURE 3.3: Twenty–six seconds of brain waves recorded during REM state.

3.3 The Influence of Circadian Rhythms

Biological variations that occur in the course of 24 hours are called circadian rhythms. Circadian rhythms are controlled by the biological clock of the body. Many bodily functions follow the biologic clock, but sleep and wakefulness comprise the most important circadian rhythm. Circadian sleep rhythm is one of the several body rhythms modulated by the hypothalamus.

Light directly affects the circadian sleep rhythm. Light is called *zeitgeber*, a German word meaning time-giver, because it sets the biological clock.

Body temperature cycles are also under control of the hypothalamus. An increase in body temperature is seen during the course of the day and a decrease is observed during the night. The temperature peaks and troughs are thought to mirror the sleep rhythm. People who are alert late in the evening (i.e. evening types) have body temperature peaks late in the evening, while those who find themselves most alert early in the morning (i.e. morning types) have body temperature peaks early in the morning.

Melatonin (a chemical produced by the pineal gland in the brain and a hormone associated with sleep,) has been implicated as a modulator of light entrainment. It is secreted maximally during the night. Prolactin, testosterone and growth hormone also demonstrate circadian rhythms, with maximal secretion during the night. Figure 3.4 shows a typical concentration of melatonin in a healthy adult man. Sleep and wakefulness are influenced by different neu-

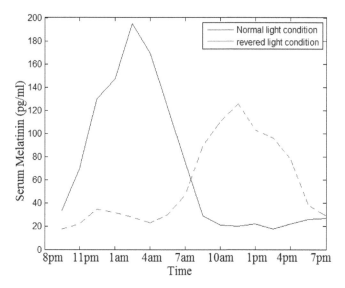

FIGURE 3.4: A typical concentration of melatonin in a healthy adult man.

rotransmitters in the brain. Some substances can change the balance of these neurotransmitters and affect our sleep and wakefulness. Caffeinated drinks (for example, coffee) and medicines (for example, diet pills) stimulate some parts of the brain and can cause difficulty in falling asleep. Many drugs prescribed for the treatment of depression suppress REM sleep.

Heavy smokers who smoke heavily often sleep very lightly and have reduced duration of REM sleep. They tend to wake up after three or four hours of sleep due to nicotine withdrawal. Some people who have insomnia may use alcohol. Even though alcohol may help people to fall into light sleep, it deprives them of REM sleep and the deeper and more restorative stages of sleep. Alcohol keeps them in the lighter stages of sleep from which they can be awakened easily. During REM sleep, we lose some of our ability to regulate our body temperature. Therefore, abnormally hot or cold temperatures can disrupt our

REM sleep. If our REM sleep is disturbed, the normal sleep cycle progression is affected during the next sleeping time, and there is a possibility of slipping directly into REM sleep and going through long periods of REM sleep until the duration of REM sleep that is lost is caught up.

Generally, sleep disruption by any cause can be a reason for an increase in seizure frequency or severity. It can also have negative effect on short-term memory, concentration, and mood. Seizure, itself, during the night can disrupt sleep. Also, using any anticonvulsant drug may affect sleep in different ways. Approximately ninety different sleep disorders, including snoring, obstructive sleep apnea hypopnea syndrome (OSAHS), insomnia, narcolepsy, bruxism, and restless leg syndrome have been reported by the International Classification of Sleep [10].

Both the frequency of seizure and the locality of seizure sources within the brain may change in different sleep stages and wakefulness.

3.4   Sleep Deprivation

Sleep deprivation is evaluated in terms of the tasks impaired and the average duration. In tasks requiring judgment, increasingly risky behaviours emerge as the total sleep duration is limited to five hours per night. The high cost of an action is seemingly ignored as the sleep-deprived person focuses on limited benefits. These findings can be explained by the fact that metabolism in the prefrontal and parietal associational areas of the brain decrease in individuals deprived of sleep for 24 hours. These brain areas are important for judgment, impulse control, attention, and visual association.

Sleep deprivation is a relative concept. Small amounts of sleep loss (for example, one hour per night over many nights) produce subtle cognitive impairment, which may go unrecognized. More severe restriction of sleep for a week leads to profound cognitive deficits, which may also go unrecognized by the individual. If one feels drowsy during the day, falls asleep for very short periods of time (5 minutes or so), or regularly falls asleep immediately after lying down, he is probably sleep-deprived.

Many studies have made it clear that sleep deprivation is dangerous. With decreased sleep, higher-order cognitive tasks are impaired early and disproportionately. On tasks used for testing coordination, sleep-deprived people perform as poorly as or worse than people who are intoxicated. Total sleep duration of seven hours per night over one week has resulted in decreased speed in tasks of both simple reaction time and more demanding computer-generated mathematical problem solving. Total sleep duration of five hours per night over one week shows both a decrease in speed and the beginning of accuracy failure.

Using sleep deprivation for detection and diagnosis of some brain abnormalities has been reported by some researchers [12, 13]. It consists of sleep loss for 24–26 hours. This was used by Klingler [14] to detect the epileptic discharges that could otherwise be missed. Based on these studies it has also been concluded that sleep depravation is a genuine activation method [15]. Its efficacy in provoking abnormal EEG discharges is not due to drowsiness. Using the information in stage III of sleep, the focal and generalized seizure may be classified [16].

In a number of experimental scenarios there are situations where the subjects are *forced to sleep* regularly for a number of days. Both physiological and biological symptoms are then measured and used to diagnose the abnormalities.

3.5   Psychological Effects

In majority of sleep measurements and studies EEG has been used in combination with a variety of other physiological parameters. EEG studies have documented abnormalities in sleep patterns in psychiatric patients with suicidal behaviour, including longer sleep latency, increased REM time and increased phasic REM activity. Sabo et al. [17] compared sleep EEG characteristics of adult depressives with and without a history of suicide attempts and noted that those who attempted suicide had consistently more REM time and phasic activity in the second REM period but less delta wave counts in the fourth non-REM period. Another study [18] conducted at the same laboratory replicated the findings with psychotic patients. On the basis of two studies [18], it has been suggested that the association between REM sleep and suicidality may cut across diagnostic boundaries and that sleep EEG changes may have a predictive value for future suicidal behaviour. REM sleep changes were later replicated by other studies in suicidal schizophrenia [19] and depression [20].

Three cross-sectional studies examined the relationship between sleep EEG and suicidality in depressed adolescents. Dahl et al. [21] compared sleep EEG between a depressed suicidal group, a depressed nonsuicidal group, and normal controls. Their results indicated that suicidal depressed patients had significantly prolonged sleep latency and increased REM phasic activity with a trend for reduced REM latency compared to both nonsuicidal depressed and control groups. Goetz et al. [22] and McCracken et al. [25] also reached the conclusion that there exists a greater REM density among depressive suicidal adolescents.

Study of normal aging and transient cognitive disorders in the elderly has also shown that the most frequent abnormality in the EEG of elderly subjects is slowing of alpha frequency whereas most healthy individuals maintain alpha activity within 9–11 Hz [23] and [24].

3.6   Detection and Monitoring of Brain Abnormalities During Sleep by EEG Analysis

EEG provides important and unique information about the sleeping brain. Polysomnography (PSG) has been the well-established method of sleep analysis and the main diagnostic tool in sleep medicine, which interprets the sleep signal macrostructure based on the criteria explained by Rechtschaffen and Kales (R&K) [26]. Polysomnography or sleep study is a multi-parametric test used in the study of sleep, and as a diagnostic tool in sleep medicine. The test result is called a polysomnogram, PSG is a comprehensive recording of the biophysiological changes occurring during sleep. It is usually performed at night. The PSG monitors many body functions including brain (EEG), eye movements (EOG), muscle activity or skeletal muscle activation (EMG) and heart rhythm (ECG) during sleep.

A comprehensive overview of the methods used for detection of various waveforms from the EEG signals has been presented in Chapter 12 of [1]. This book also provides the results of applications of the corresponding tools and algorithms. These methods include blind source separation (BSS), matching pursuit, time-frequency domain using wavelet transform, higher-order statistics, artificial neural networks, and also model-based approaches. Model-based approaches rely on an a priori knowledge about the mechanism of generation of the data or the data itself [1]. Characterizing a physiological signal generation model for NREM has also been under study by several researchers [28, 29]. These models describe how the depth of NREM sleep is related to the neuronal mechanism that generates slow waves. This mechanism is essentially feedback through closed loops in neuronal networks or through the interplay between ion currents in single cells. It is established that the depth of NREM sleep modulates the gain of the feedback loops [27]. According to this model, the sleep-related variations in the slow wave power (SWP) result from variations in the feedback gain. Therefore, increasing depth of sleep is related to an increasing gain in the neuronal feedback loops that generates the low frequency EEG.

In a number of applications a hybrid of the above methods have been used to exploit multimodal physiological data recordings. This is certainly because diagnosis of sleep disorders and other related abnormalities may not be complete unless other physiological symptoms are studied. These symptoms manifest themselves within other physiological and pneumological signals such as respiratory airflow, position of the patients, electromyogram (EMG) signal, hypnogram, level of SaO2, abdominal effort, and thoracic effort which may also be considered in the classification system.

Time-frequency methods have been more robust and popular and used as a benchmark for comparing with other algorithms. In the following section an adaptive SSA method is proposed to better estimate the stages of sleep (so called sleep scoring) from the EEG signals.

3.7   SSA for Sleep Scoring

Here, a new supervised approach for decomposition of single channel signal mixtures is introduced. It is seen that the grouping stage is governed by another frequency-based approach. In addition, it is demonstrated that the performance of the traditional SSA algorithm is significantly improved by applying tensor decomposition instead of traditional singular value decomposition (SVD). In this subspace analysis method, the inherent frequency diversity of the sleep EEG data has been effectively exploited to highlight the subspace of interest [31].

3.7.1   Methodology

In analysis and processing of biomedical signals, often some mixtures of the source signals are recorded. Hence, our task is to decompose them into their constituent components and retrieve the underlying sources. For instance, in the analysis of EEG signals, it is required to extract neurophysiologically meaningful information in applications characterizing event related potentials, such as brain computer interfacing (BCI), seizure detection, and sleep analysis.

In multichannel data, this problem is efficiently handled by employing BSS techniques, which unmix the given signal mixtures into their constituent sources [1]. BSS is applied when the number of sources is equal or less than the number of electrodes. It fails for single channel, or generally underdetermined, recordings. However, there are several applications where just one channel is used. For example, restoring an EMG signal contaminated by an ECG artefact, in single channel deep brain recordings, where the neuronal spikes are not easily separable from noise [6], and different sleep stages with various dominant frequency bands where each frequency band is prevailing in a special recorded channel.

As explored in some of the chapters of this book, recently SSA has been employed in biomedical signal processing applications such as separation of EMG and ECG [32] and restoring lung sound from heart sound [33]. A supervised SSA has been attempted in [34] to detect spikes from noisy signals. This method, however, does not exploit the narrow band property of cyclic data. In EEG, the brain rhythms manifest themselves as narrow frequency band components well distinct from each other.

As a good example, sleep is a dynamic process which consists of different stages with different neural activity levels. Each stage is characterised by a distinct set of physiological and neurological features and a dominant frequency band. Therefore, sleep study can be considered as one good application of this method.

In development of this method a tensor-based approach is suggested which better deals with nonstationarity of most of physiological signals, particularly EEG data. Sleep EEG has a nonstationary structure since its statistical properties in both time and frequency change during the stages of sleep. Furthermore, in SSA, selecting the desired subgroup of eigenvalues is still an open problem.

As the grouping procedure plays an important role in the reconstruction stage, the importance of developing a proper subspace grouping is very obvious here. Hence, the objective is to extend the SSA algorithm and theory to separate the single channel data more effectively.

In an attempt to exploit the nonstationarity of the signals, the 1D arrays are converted to 3D arrays, and therefore the decomposition stage of the SSA problem is converted to a tensor factorization problem [35]. Tensor representation is a way to algebraically describe the datasets by preserving their multiway structures. Tensor factorization methods can extract the true underlying structures of the data. Before we move on any further in application, let us briefly look at the tensor-based SSA.

3.7.2 Tensor-Based SSA

Application of SSA to real data does not exploit the inherent nonstationarity of the biomedical signals and therefore may fail in actual data decomposition. Tensor-based SSA (TSSA) is a robust solution to this problem. Just as in SSA, the first stage of TSSA includes an embedding operation followed by a tensor decomposition method instead of SVD. In the embedding stage a 1D time series $Y_N = (y_1, \ldots, y_N)$ with length N is mapped into tensor $\mathbf{X}$. To do that, first Y_N is segmented using non-overlapping window of size L and a $\lfloor N/L \rfloor \times L$ matrix $\mathbf{X}_t$ is obtained from Y_N:

$$
\mathbf{X}_t = \begin{pmatrix}
y_1 & y_2 & \cdots & y_L \\
y_{L+1} & y_{L+2} & \cdots & y_{2L} \\
\vdots & \vdots & \ddots & \vdots \\
y_{(I-1)L} & y_{(I-1)L+1} & \cdots & y_{IL}
\end{pmatrix}
\tag{3.1}
$$

where, $I = \lfloor N/L \rfloor$. This is a much larger window compared to the SSA embedding dimension but small enough to avoid nonstationarity. Then, this matrix is converted to tensor $\mathbf{X}$ as demonstrated in Figure 3.5 by considering each slab of the tensor as a windowed version of $\mathbf{X}_t$. In this work, the segmentation is performed in one direction, but it can be extended to move the window in both directions. The matrix to tensor conversion can be mathematically

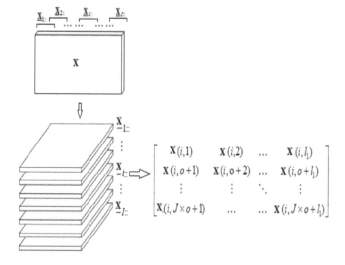

FIGURE 3.5: The matrix $\mathbf{X}$ is converted to tensor $\underline{\mathbf{X}}$ where J is the number of segments obtained by window size l_1 and overlapping interval $L_1 - o$.

explained by the following equation:

$$
\begin{aligned}
\underline{\mathbf{X}}_{i::} &= \mathbf{X}_t(:, (j-1)o+1 : (j-1)o+l_1) \\
j &= 1, 2, ..., J, J = \lfloor (L - L_1)/o + 1 \rfloor \\
i &= 1, 2, ..., I,
\end{aligned}
\tag{3.2}
$$

where $L_1 - o$ is the overlapping size between the successive windows and L_1 is the window size. At this stage, we have a 3D tensor to be decomposed. Parallel factor analysis (PARAFAC) is a canonical decomposition algorithm that can be considered as a generalization of bilinear principal component analysis (PCA) [36, 37]. The fundamental expression of the PARAFAC technique is given as [38]–[40]:

$$
x_{ijk} = \sum_{r=1}^{R} a_{ir} b_{jr} c_{kr} + e_{ijk},
\tag{3.3}
$$

where x_{ijk} is the (i, j, k)th element in the three-way dataset, R is the number of common components or model order, a_{ir}, b_{jr} and c_{kr} are the elements in $\mathbf{A}$, $\mathbf{B}$ and $\mathbf{C}$ respectively, and e_{ijk} is the residual term. Many adaptive algorithms studied in adaptive filtering context can be applied in an alternating fashion to solve this problem. Here the alternating least squares (ALS) optimization method is used to fit the PARAFAC model [41]. In this convex approach iteratively, B and C are fixed to solve for $\mathbf{A}$; then, $\mathbf{A}$ and $\mathbf{C}$ are fixed to solve for $\mathbf{B}$, and next $\mathbf{A}$ and $\mathbf{B}$ are fixed to solve for $\mathbf{C}$ in an alternating manner until reaching some convergence. One important characteristic of PARAFAC is that it is convex, and performs well in certain underdetermined cases where the number of sources is comparable with tensor rank [42]. Moreover, by segmenting and decomposing the signal in this way, the length of signal being processed at one time can increase.

After this stage, m disjoint subsets of indices I_k are specified. Selecting a proper subgroup of components has a significant impact on the final result. If I_{ks} for $k = 1, \ldots, m$ indicate the group labels, we have:

$$I_1 \cup I_2 \cup \ldots \cup I_m \ = \ \{1, 2, \ldots, R\}$$

$$\underline{\mathbf{X}} \ = \ \sum_{i=1}^{m} \underline{\mathbf{X}}_{I_i} \tag{3.4}$$

$$\underline{\mathbf{X}}_K \ = \ \sum_{k \in K} \mathbf{A}_k \circ \mathbf{B}_k \circ \mathbf{C}_k,$$

where $\circ$ denotes outer product and $\underline{\mathbf{X}}$ indicates the tensor obtained from I_kth subgroup. Finally, a reconstruction of the original signal can be obtained using a hankelisation procedure which is performed by converting the signal segments to a Hankel matrix:

$$\hat{\mathbf{X}}_{I_i::} = H\mathbf{X}_{I_i::} \tag{3.5}$$

Then, the data is reconstructed by one-to-one correspondence. Hereafter, the grouping stage is an important problem. However, as stated before, frequency diversity of the information of interest within the sleep data is the main a priori to select the appropriate groups of eigentriples. In the following session this property is nicely exploited by means of empirical mode decomposition (EMD). Therefore, the overall system will be a hybrid of tensor-based SSA and EMD, namely TSSA-EMD.

3.7.3 EMD

EMD is a nonlinear technique in signal decomposition for adaptively representing nonstationary signals as sums of zero-mean amplitude or frequency modulated signals [30]. To better understand the method a brief overview of EMD is provided in this section.

Similar to the very popular Fourier series, to perform EMD the signal oscillations at a very local level is considered. By looking at the evolution of a signal y_t between two consecutive extrema (say, two minima occurring at times t_- and t_+), we can heuristically define a (local) high-frequency part ($d_t, t_- \leq t \leq t_+$), or local detail, which corresponds to the oscillation terminating at the two minima and passing through the maximum which necessarily exists in between them. For the picture to be complete, one still has to identify the corresponding (local) low-frequency part m_t, or local trend, so that we have $y_t = m_t + d_t$ for $t_- \leq t \leq t_+$. Assuming that this is done in some proper way for all the constituent oscillations within the entire signal, the procedure can then be applied to the residual consisting of all local trends, and constitutive components of a signal can therefore be iteratively extracted. Therefore, given a signal y_t, the EMD algorithm can be summarised as follows:

1. identify all extrema of y_t,

2. interpolate between minima, ending up with some envelope e_{min_t}, and between maxima, ending up with some envelope e_{max_t},

3. compute the mean $m_t = (e_{min_t} + e_{max_t})/2$,

4. extract the detail $d_t = y_t - m_t$,

5. iterate on the residual m_t in the same fashion as for y_t.

In practice, the above procedure has to be refined by a sifting process [30] which amounts to first iterating steps 1 to 5 upon the detail signal d_t, until this latter can be considered a zero-mean according to some stopping criterion. Once this is achieved, the detail is referred to as an intrinsic mode function (IMF), the corresponding residual is computed and Step 5 applies. Within the construction process, the number of extrema is decreased when going from one residual to the next, and the whole decomposition is guaranteed to be completed with a finite number of modes. Figure 3.6 refers to the signal decomposition using EMD.

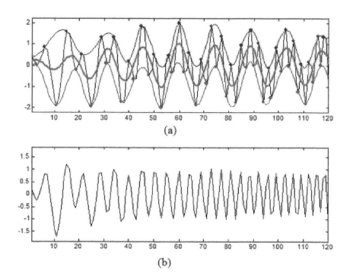

FIGURE 3.6: EMD process; (a) the original signals, maxima and minima envelopes and the first IMF (middle signal), and (b) residual signal which is equal to subtraction of the first IMF from the original signal. The process can continue by considering the residual as the original signal in the next stage.

The extraction of a mode is considered successful when the *sifting* process is terminated. Two conditions are to be fulfilled in this respect [30]: firstly, the number of extrema and the number of zero-crossings must differ at most by 1; secondly, the mean between the upper and lower envelopes must be less than a threshold level defined based on some criterion.

The IMFs each have a mode and in an ideal case each mode represents an oscillation component of the signal in a particular frequency. However, often there are some variations, which may be considered as noise, and even a change in frequency of the mode (often called as mode change).

3.7.4   TSSA-EMD

For the sleep data the stages of sleep are characterised by variations of normal brain rhythms particularly alpha (8–13 Hz) and delta, or slow wave, which is below 4 Hz. On the other hand, the grouping process has a significant impact on the final results. The groups are often characterised by their statistical or physical properties. In application to brain signals, the brain rhythms are well characterised by their frequency properties. As an elegant approach for exploiting the frequency diversity of the data, EMD is used here to supervise the TSSA-based subspace decomposition. EMD provides a good way to identify the number of frequency components within each subspace. So, EMD is used to select the subgroup of the desired signal, as can be seen in Figure 3.7 In other words, a number of IMFs falling within a particular frequency band are selected according to some pre-set criteria. For instance, the components that have maximum power in the desired frequency bands can be selected and used to more accurately group the eigentriples. Then, by assuming the Hankel matrix of this signal as $\mathbf{H}_{ij}$ for each segment k, the correct eigenvalue group is selected by minimizing the following objective function:

$$J(w_{ijk}) = \|h_{ijk} - \sum_{r=1}^{R} w_{ijk} a_{ir} b_{jr} c_{kr}\|^2, \tag{3.6}$$

where $\|.\|$ denotes the Frobenius norm, R is the number of common components, a_{ir}, b_{jr} and c_{kr} are the tensor factors of the PARAFAC procedure, and $W = w_{ijk}$ is a superdiagonal tensor of adaptive weights. As the objective function is convex (proof can be seen in [31]), it can be minimised using different methods. Here, a subgradient method which is an iterative method for solving convex optimisation problems is employed to obtain the optimal value of w_{ijk}:

$$w_{ijk}^{m+1} = w_{ijk}^m - \mu \nabla_{w_{ijk}} J(w_{ijk})$$

$$\nabla_{w_{ijk}} J(w_{ijk}) = -2(h_{ijk} - \sum_{r=1}^{R} w_{ijk} a_{ir} b_{jr} c_{kr}) \sum_{r=1}^{R} a_{ir} b_{jr} c_{kr},$$

where μ indicates the step size.

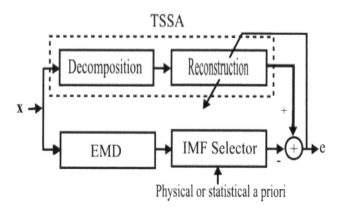

FIGURE 3.7: Block diagram of the single channel source separation system using an adaptive procedure for selecting the desired subspace. This is carried out by tuning a set of weights governed by the EMD process.

3.7.5   Application to Sleep EEG

As previously stated, sleep is a highly complex state which is also a sensitive indicator of changes in brain function such as occurs in many psychiatric and neurological conditions. Sleep is characterised by a reduction in body movements, reduced responsiveness to external stimuli, and changes in metabolic rate [1]. Hence, it is a state of unconsciousness from which a person can be aroused. Brain states during sleep and wakefulness are generated through interactions of activating and inhibiting systems within the brain. There are two distinct states with different levels of neuronal activity: non-rapid eye movement (NREM) and REM sleep. Each stage has a distinct set of physiological and neurological features, and also has a dominant frequency band [1]. Most prominently, the alpha rhythm (8–13Hz) is attenuated and delta (up to 4Hz) waves evolve as the NREM sleep deepens. Other features of NREM sleep are sleep spindles and K complexes. It has been assumed that K complexes are due to continuous sensory activation and it can be considered as a building block of slow wave (SW) sleep which has more power than the usual delta waves [39]. As the manual (by visualising) sleep scoring is a time–consuming process, automatic sleep staging methods hold promise in diagnosing alterations in the sleep process and the sleep EEG more efficiently.

At the University of Surrey Sleep Centre thirty-six healthy men and women each participated in two laboratory sessions, one involving a sleep extension protocol and the other a sleep restriction protocol. During each session polysomnography (PSG) measurements were recorded at a sampling rate of 256Hz for a baseline (BL) night (8 hours), seven condition nights, sleep extension (ES), 10 hours; sleep restriction (SR), 6 hours and a recovery night (12 hours) following a period of total sleep deprivation. In this paper, the proposed method and also two commonly used methods for EEG sleep processing, Morlet wavelet and power spectrum analysis using fast Fourier transform (FFT), were applied to sleep data to extract different frequency bands (alpha and delta). Delta rhythm is a slow brain wave which tends to have its highest amplitude during deep sleep in adults and is usually prominent frontally. Alpha can be seen in the posterior regions of the head on both sides, and emerges with closing of the eyes and relaxation. Therefore, only one channel for which more variations in alpha, and another one with more variations in delta, were chosen and after applying the following method the average power was determined for each 2s time segment. Thus, each point i in Figures 3.8 and 3.9 corresponds to ith 2s frame of the signal starting at $i \times 2 \times 256$ sample in the original signal. The results of these methods for one subject in a BL night are depicted in Figure 3.8. In addition, the result of applying the proposed method to BL, SE, and SR nights is seen in Figure 3.9. The results for other available signals follow quite the same structure for different methods. Table 3.1 indicates the start of second and SW stages according to a precise hand scoring and the proposed method by a simple thresholding on the resulted values.

In each figure the left plot indicates the changes in delta power and the right plot shows the same for the alpha band.

TABLE 3.1: Comparison between manual and proposed method in identification of the start points of second and SW stages of sleep for various types of sleep.

	Scoring			
	Manual		TF-SSA	
Condition	Second	SW	Second	SW
baseline	660	990	638	980
SR	150	390	120	380
SE	1350	1650	1321	1600

3.8 Concluding Remarks

Study of sleep EEG has opened a new direction to investigate the psychology of a human being and various sleep disorders. There is however, much more to discover from these signals. Various physiological and mental brain disorders manifest themselves differently in the sleep EEG signals. Sleep scoring is a major tool to provide more insight into the brain state of those suffering from both physical and mental abnormalities. SSA provides a robust solution to this problem.

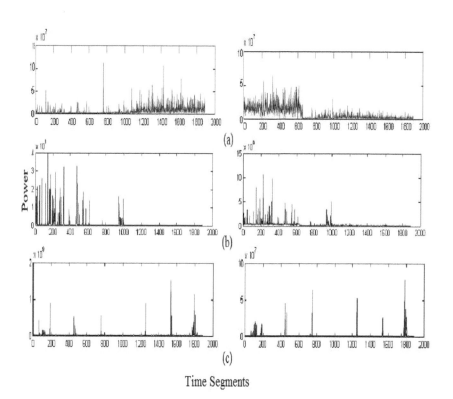

FIGURE 3.8: Changes in alpha and delta bands during one cycle of sleep for a baseline night with respect to time segments obtained using TSSA-EMD, wavelet, and FFT; (a) TSSA-EMD, (b) wavelet, and (c) FFT. In each figure the left plot indicates the changes in delta power and the right plot shows the power changes for the alpha band.

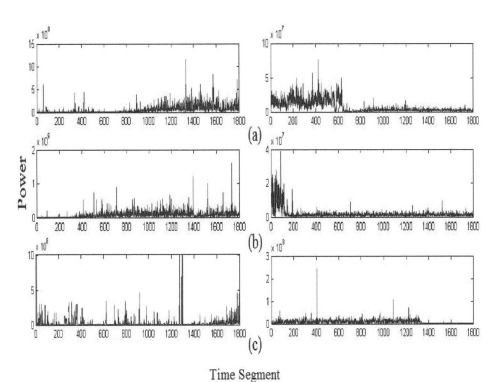

FIGURE 3.9: Changes in the alpha and delta bands during one cycle of sleep for SR, SE, and BL with respect to time segments obtained using TSSA-EMD; (a) BL night, (b) SR, and (c) SE.

As indicated in Figure 3.8, the TSSA-EMD method can determine the transitions between the stages of sleep by more accurately evaluating the alpha and delta (SW) brain activity variations. This significant achievement complies with an accurate manual scoring sleep data by clinical experts, which can be seen in Table II. Moreover, the agreement between the scorers was determined with intraclass correlation coefficients (ICC) which were computed using a 2-way analysis of variance [43]. The higher numbers in ICC represent better agreement between raters. We obtained ICC= 0.99 which shows almost complete agreement. The proposed method performs well for detecting the stage changes. Wavelet transform performs satisfactorily for detecting the start of the first stage. However, it cannot show the deep sleep stage very well, according to Figure 3.8(b). Simple filtering using FFT also could not reveal the necessary information about sleep stages (Figure 3.8(c)). Therefore, based on the illustrated results, the proposed method performs better than the conventional methods, i.e. WT and FFT, for sleep processing.

Another important result of our method is to extract the K-complexes from sleep signals. K complexes are seen in EEG sleep data after finishing the first stage of sleep which have the same frequency band as delta but with significantly higher amplitude. Therefore, as can be seen in Figure 3.8, the sudden increases can show the locations of sudden changes in sleep. Figure 3.9 also confirms the results of some previous experiments in which a fast transition from alpha to slow waves is shown in the SR condition and the rivers in the SE condition.

Detection and classification of mental diseases from the sleep EEG signals, however, require more deep analysis of the data by developing and utilizing advanced digital signal processing techniques. The analyses become more challenging when other parameters such as age are involved. For example, in neonates many different types of complex waveforms may be observed for which the origin and causes are still unknown. On the other hand, there are some similarities between the normal rhythms within the sleep EEG signals and the EEGs of abnormal rhythms such as epileptic seizure and hyperventilation. An efficient algorithm (based on solely EEG or combined with other physiological signals) should be able to differentiate between such different brain states.

Bibliography

[1] Sanei, S. (2013). *Adaptive Processing of Brain Signals*. John Wiley & Sons.

[2] Lavie, P. (1996). *The Enchanted World of Sleep*, Yale University Press.

[3] Steriade, M. (1992). Basic mechanisms of sleep generation. *Neurology*, **42**(6), 9–17.

[4] Kubicki, S., Scheuler, W. and Wittenbecher, H. (1991). Short-term sleep EEG recordings after partial sleep deprivation as a routine procedure in order to uncover epileptic phenomena: an evaluation of 719 EEG recordings. *Epilepsy Research*, Supplement **2**, 217–230.

[5] Niedermeyer, E. (1999). *Sleep and EEG*, Chapter 10 In *Electroencephalography Basic Principles, Clinical Applications, and Related Fields*, Eds. E. Niedermeyer and F. L. Da Silva, 174–188.

[6] Koshino, Y., Nishio, M., Murata, T., Omori, M., Murata, I., Sakamoto, M., and Isaki, K. (1993). The influence of light drowsiness on the latency and amplitude of P300. *Clinical EEG and Neuroscience*, **24**(3), 110–113.

[7] Wada, Y., Nanbu, Y., Koshino, Y., Shimada, Y., and Hashimoto, T. (1996). Inter- and intrahemispheric EEG coherence during light drowsiness. *Clinical EEG and Neuroscience*, **27**(2), 24–88.

[8] Niedermeyer, E. (1999). Maturation of EEG: development of walking and sleep patterns. In *Chap. 11, Electroencephalography*, Eds. E. Niedermeyer, and F. Lopez de silva, Williams and Wilkins.

[9] Bonanni, E., Di Coscio, E., Maestri, M., Carnicelli, L., Tsekou, H., Economou, N. T., Paparrigopoulos, T., Bonakis, A., Papageorgiou, S. G., Vassilopoulos, D., Soldatos, C. R., Murri, L., Ktonas, P. Y. (2012). Differences in EEG delta frequency characteristics and patterns in slow-wave sleep between dementia patients and controls: a pilot study. *Clinical Neurophysiology*, **29**(1), 50–54.

[10] Shiomi, F. K., Pisa, I. T., de Campos, C. J. R. (2011). Computerized analysis of snoring in sleep apnea syndrome. *Brazilian Journal of Otorhinolaryngology*, **77**(4), 488–498.

[11] Brzezinski, A. (1997). Melatonin in Humans. *New England Journal of Medicine*, **336**(3), 186–195.

[12] Cabrero-Canosa, M., Hernandez-Pereira, E., and Moret-Bonillo, V. (2004). Intelligent dignosis of sleep apnea syndrome. *IEEE: Engineering in Medicine and Biology Magazine*, **23**(2), 72–81.

[13] Redmond, S. J. and Heneghan, C. (2006). Cardiorespiratory-based sleep staging in subjects with obstructive sleep apnea. *Biomedical Engineering, IEEE Transactions*, **51**(3), 485–496.

[14] Klingler, D., Trägner, H., and Deisenhammer, E. (1990). The nature of the influence of sleep deprivation on the EEG. *Epilepsy research*, Supplement **2**, 231–234.

[15] Jovanović, U. J. (1991). General considerations of sleep and sleep deprivation. *Epilepsy Research*, Supplement **2**, 205–215.

[16] Naitoh, P., Kelly, T. L., Englund, C. (1990). Health effects of sleep deprivation. *Occupational Medicine*, **5**(2), 209–37.

[17] Sabo, E., Reynolds, C. F., Kupfer, D. J., and Berman, S. R. (1991). Sleep, depression, and suicide. *Psychiatry Research*, **36**(3), 265–277.

[18] Weitzenblum, E., and Racineux, J. L. (2004). *Syndrome dápnées obstructives du sommeil*. 2nd Ed., Masson Press, Paris, France.

[19] Man, G. C., and Kang, B. V. (1995). Validation of a portable sleep apnea monitoring device. *CHEST Journal*, **108**(2), 388–393.

[20] Porée, F., Kachenoura, A., Gauvrit, H., Morvan, C., Carrault, G., and Senhadji, L. (2006). Blind source separation for ambulatory sleep recording. *Information Technology in Biomedicine, IEEE Transactions*, **10**(2), 293–301.

[21] Dahl, R. E., Puig-Antich, J., Ryan, N. D., Nelson, B., Dachille, S., Cunningham, S. L., and Klepper, T. P. (1990). EEG sleep in adolescents with major depression: the role of suicidality and inpatient status. *Journal of Affective Disorders*, **19**(1), 63–75.

[22] Goetz, R. R., Puig-Antich, J., Dahl, R. E., Ryan, N. D., Asnis, G. M., and Nelson, B. (1991). EEG sleep of young adults with major depression: a controlled study. *Journal of Affective Disorders*, **22**(1), 91–100.

[23] Van Sweden, B., Wauquier, A., and Niedermeyer, E. (1999). Normal aging and transient cognitive disorders in the elderly. *Electroencephalography: basic principles, clinical applications, and related fields*, Baltimore: Lippincott Williams & Wilkins, 340–348.

[24] Klass, D. W., and Brenner, R. P. (1995). Electroencephalography of the elderly. *Journal of Clinical Neurophysiology*, **12**(2), 116–131.

[25] McCracken, J. T., Poland, R. E., Lutchmansingh, P., and Edwards, C. (1997). Sleep electroencephalographic abnormalities in adolescent depressives: effects of scopolamine. *Biological Psychiatry*, **42**(7), 577–584.

[26] Rechtschaffen, A., and Kales, A. (1968). A manual of standardized terminology and scoring system for sleep stages of human subjects. Los Angeles: Brain Information Service. *Brain Research Institute, University of California at Los Angeles.* Ed. Ser. National Institutes of Health publications. Washington DC: U.S. Government Printing Office, 204.

[27] Steriade, M., McCormick, D. A., and Sejnowski, T. J. (1993). Thalamo-cortical oscillations in the sleeping and aroused brain. *Science*, **262**(5134), 679–685.

[28] Kemp, B., Zwinderman, A. H., Tuk, B., Kamphuisen, H. A., and Oberye, J. J. (2000). Analysis of a sleep-dependent neuronal feedback loop: the slow-wave microcontinuity of the EEG. *Biomedical Engineering, IEEE Transactions*, **47**(9), 1185–1194.

[29] Merica, H., and Fortune, R. D. (2003). A unique pattern of sleep structure is found to be identical at all cortical sites: a neurobiological interpretation. *Cerebral Cortex*, **13**(10), 1044–1050.

[30] Huang, N. E., Shen, Z., Long, S. R., Wu, M. C., Shih, H. H., Zheng, Q., and Liu, H. H. (1998). The empirical mode decomposition and the Hilbert spectrum for nonlinear and non-stationary time series analysis. *In Proceedings of the Royal Society of London A: Mathematical, Physical and Engineering Sciences*, **454**(1971), 903-995. The Royal Society.

[31] Kouchaki, S., Sanei, S., and Dijk, D. yy⁻J. (2013). Single Channel Source Separation using Tensor based Singular Spectrum Analysis for Automatic Scoring of Sleep EEG. *IEEE Transactions on Biomedical Signal Processing*, under review.

[32] Sanei, S., Lee, T. K. M., and Abolghasemi, V. (2012). A new adaptive line enhancer based on singular spectrum analysis. *Biomedical Engineering, IEEE Transactions*, **59**(2), 428–434.

[33] Ghaderi, F., Mohseni, H. R., and Sanei, S. (2011). Localizing heart sounds in respiratory signals using singular spectrum analysis. *Biomedical Engineering, IEEE Transactions*, **58**(12), 3360–3367.

[34] Sanei, S., Ghodsi, M., and Hassani, H. (2011). An adaptive singular spectrum analysis approach to murmur detection from heart sounds. *Medical Engineering and Physics*, **33**(3), 362–367.

[35] Kouchaki, S., Sanei, S., Arbon, E. L., Dijk, D. -J. (2015). Tensor based Singular Spectrum Analysis for Automatic Scoring of Sleep EEG. *IEEE Transactions on neural systems and rehabilitation engineering: a publication of the IEEE Engineering in Medicine and Biology Society*, **23**(1), 1–9.

[36] Harshman, R. A. (1970). Foundations of the PARAFAC procedure: Model and conditions for an explanatory multi-mode factor analysis. *UCLA Working Papers in phonetics*, **16**, 1–84.

[37] Carroll, J. D., and Chang, J. J. (1970). Analysis of individual differences in multidimensional scaling via an N-way generalization of "Eckart-Young" decomposition. *Psychometrika*, **35**(3), 283–319.

[38] Harshman, R. A., and Berenbaum, S. A. (1981). Basic concepts underlying the PARAFAC-CANDECOMP three-way factor analysis model and its application to longitudinal data. *Present and past in middle life*, Academic Press, NY, 435–459.

[39] Halász, P., Pal, I., and Rajna, P. (1984). K-complex formation of the EEG in sleep. A survey and new examinations. *Acta Physiologica Hungarica*, **65**(1), 3–35.

[40] Burdick, D. S. (1995). An introduction to tensor products with applications to multiway data analysis. *Chemometrics and Intelligent Laboratory Systems*, **28**(2), 229–237.

[41] Sands, R., and Young, F. W. (1980). Component models for three-way data: An alternating least squares algorithm with optimal scaling features. *Psychometrika*, **45**(1), 39–67.

[42] De Lathauwer, L., and Castaing, J. (2008). Blind identification of underdetermined mixtures by simultaneous matrix diagonalization. *Signal Processing, IEEE Transactions*, **56**(3), 1096–1105.

[43] Shrout, P. E., and Fleiss, J. L. (1979). Intraclass correlations: uses in assessing rater reliability. *Psychological bulletin*, **86**(2), 420-428.

Chapter 4

Adaptive SSA and its Application to Biomedical Source Separation

4.1   Introduction

Selection of eigentriples for efficient separation or decomposition of signals often cannot be achieved if no a priori information is available. Such information stems from diversities of the original sources or medium. This kind of information can nicely be exploited in the design of an adaptive SSA system. The behaviour of such systems is quite similar to those of conventional adaptive filters [1]. Adaptive filters are the basic structure for many adaptive and optimisation systems which broadly exist and are used for modelling natural and physical systems and plants. These systems are described by a set of parameters which can be estimated using a number of algorithms. The established learning algorithms for the estimation of adaptive filter parameters are meant for linear systems. However, there are attempts to tackle solving such problems for nonlinear systems too. Figure 4.1 shows the block diagram of a simple adaptive filter to model a linear system.

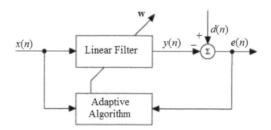

FIGURE 4.1: A conventional adaptive filter.

In such systems the vector of filter coefficients **w** is (often iteratively) estimated in order to minimise the error $e(n)$ between the filter output $y(n)$ and a desired target signal $d(n)$. As can be viewed in this figure, it is essential to have access to the target signal $d(n)$. After the filter is designed, it can be used as the model for the environment or the physical system, or it can be used as a classifier to distinguish between the right and wrong inputs.

The adaptive filters and their corresponding optimisation methods are not however limited to the above conventional form and also the solutions provided for that [1]. The nature of the input (sparse, periodic, nonstationary, ...) and nonlinearity of the physical systems (time variant, non-homogenous, ...) to be modelled often restrict the types of solutions and the optimisation approaches.

Adaptive filters have tremendous number of applications from adaptive noise cancellation [2] to multi-channel communication receiver design. In this chapter this concept is extended to adaptive SSA and two means of adaptivity for SSA are reviewed.

4.2   Adaptive Line Enhancement

Most of the signals recorded from the human body are periodic, quasi-periodic, or cyclostationary (i.e. some statistics of the data are periodic) since they originate from the sources which contribute to human body's cycle. These signals are often buried in noise or mixed with other periodic or non-periodic signals. Extraction of such cyclic activities is very important for monitoring the status of the patients.

One of the well-known applications of adaptive filters is in restoring periodic signals from noise. Such filters are called adaptive line enhancers (ALE). ALE has been widely used for separation of a low-level sinusoid or narrow-band signal from broad-band noise. This has been a classic problem in the field of signal processing, and was initially introduced for this purpose by Widrow et al. [2]. The general block diagram of the ALE is depicted in Figure 4.2.

The ALE input $s(t)$ is assumed to be the sum of a narrow-band signal $x(t)$, and a broad-band signal $n(t)$. The parameters of the prediction filter $\mathbf{w}$ are adapted in such a way that the statistical mean squared error (MSE), $E[e^2(t)]$, where $E[.]$ stands for expectation, is minimised. The ALE operates by virtue of the difference between the correlation lengths of $x(t)$ and $n(t)$. The delay parameter Δ should be chosen larger than the correlation length of $n(t)$, but smaller than the correlation length of $x(t)$. In this case, it is possible for $\mathbf{w}$ to make a $\Delta - step$ ahead prediction of $x(t - \Delta)$ based on the present and past samples of $s(t - \Delta)$. However, $\mathbf{w}$ is not able to predict $n(t)$ from knowledge about present and past samples of $n(t - \Delta)$. As a result, after the parameters of $\mathbf{w}$ have converged towards their optimal values, the error signal $e(t)$ is approximately equal to $n(t)$ and the ALE output $\hat{x}(t)$ is approximately equal to $x(t)$.

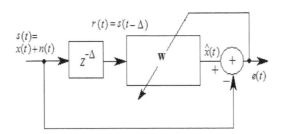

FIGURE 4.2: Traditional adaptive line enhancer.

The ALE has applications in many areas such as communications, sound, and biomedical signal processing. Although ALE is an effective tool for single channel signal denoising, it uses order statistics of the data within its minimisation criterion. On the other hand, its application is limited to having narrow band signals in the input and white noise in the output (white noise refers to temporally and statistically uncorrelated noise of a flat spectrum). Therefore, its application is limited in the cases where the signal is wideband and the noise is not white.

In this chapter the aim is to exploit the capability of SSA in introducing a new enhancer similar to ALE which can be applied to periodic signals with wider frequency bands, disregarding the noise statistics. SSA can decompose the signal space into distinct subspaces. This concept is exploited in designing a new high performance ALE.

4.3   SSA-Based ALE and its Application to Separation of ECG from Recorded EMG

In order to apply SSA for separation or denoising the signals, the corresponding subspace of the desired signal should be identified. This is probably the main shortcoming of basic SSA for single channel signal separation or denoising. In some very recent applications such as [6, 10], [8]–[10] selection of periodic components has been decided by clustering the eigentriples [10] or establishing some criteria [8]–[10]. Such criteria can be set only if the periodic signal is narrow band or well defined. Even then, the signal may not be fully reconstructed and some error is often involved.

For the case of periodic signals, given the signal period, the information can be used in a way similar to that used in ALE. A delayed (by one period) version of the signal is then used as a reference for adaptive reconstruction of the signal. Unlike for basic SSA, where the eigenvalues of the desired signals have to be selected manually, here, the algorithm is adaptive and the signal periodicity is fully exploited (as for the ALE). A block diagram of the overall SSA-based ALE is depicted in Figure 4.3.

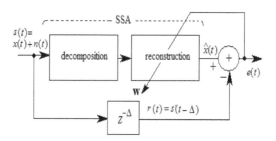

FIGURE 4.3: Block diagram of the proposed SSA-based ALE.

Assume the Hankel matrix for $r(t) = s(t - \Delta)$ is $\mathbf{R}$. During reconstruction the following cost function can be minimised in order to make sure the correct subgroup of eigentriples is selected for reconstruction of the periodic signal.

$$J(\mathbf{W}) = \|\mathbf{R} - \mathbf{U}\mathbf{W}\mathbf{\Lambda}^{1/2}\mathbf{V}^T\|_F^2, \tag{4.1}$$

where $\|.\|_F$ denotes Frobenius norm. $\mathbf{U}$, $\mathbf{\Lambda}$ (diagonal matrix of eigenvalues), and $\mathbf{V}$ are the SVD factors (derived in the SVD stage of SSA) and $\mathbf{W}$ is a $d \times d$ diagonal matrix of adaptive weights w_{ij}. Its size is the same as $\mathbf{\Lambda}$'s. To minimise this cost function different recursive optimisation methods can be used. A simple gradient approach, often used in a gradient descent approach for designing the adaptive filters, leads to the following update equation:

$$\mathbf{W}_{k+1} = \mathbf{W}_k - \mu \mathbf{U}\mathbf{\Lambda}^{1/2}\mathbf{V}^T(\mathbf{R} - \mathbf{W}_k\mathbf{\Lambda}^{1/2}\mathbf{V}^T)^T, \tag{4.2}$$

where μ is the iteration step size (which is often set manually but can be adapted to the convergence rate or made proportional to the number of iterations). The weights of the filter, $\mathbf{W}$, are estimated using (2). In the reconstruction process $\mathbf{W}$ is multiplied by $\mathbf{\Lambda}^{1/2}$ and the desired signal is recovered during the SSA reconstruction process.

In this application the periodic nature of the source signals has been effectively exploited. No prior assumption about the correspondence between the eigenvalues is considered during the optimisation process. In addition, the optimisation is entirely based on a similarity measure between the signals and their shifted version and is less dependent on the nature of the error.

In such optimisation processes, further improvements can be achieved if more information about the sources or their mixing systems is available. Such information may impose some sorts of constraint on the solution and can better guide the direction of optimisation. One such a priori is the frequency diversity stating that the desired source signals fall within certain frequency bands. In the chapter dedicated to analysis of sleep signals, this problem has been solved adaptively using joint EMD-SSA analysis. This approach will be revisited in the later sections of this chapter.

Another one of such a priori information in the case of periodic signals, is that the eigenspectrum is sparse since periodic signals contribute to a very small number of nonzero eigenvalues. In the following section this information is exploited within the optimisation process to calculate $\mathbf{W}$.

4.4   Incorporating Sparsity Constraint

The weighted eigenvalue pattern for periodic signals is expected to have a small number of non-zero components. This concept can be exploited and used to further improve the optimization algorithm. To maximise the sparsity, the number of non-zero components (weighted eigenvalues) should be minimised. This is equivalent to minimising the L_0-norm of $\mathbf{W}\mathbf{\Lambda}^{1/2}$. Since this quantity is NP hard, it is often approximated by the L_1-norm. The two costs can be combined to change the above constrained problem into an unconstrained one using a penalty term. Therefore, the overall cost function can be written as:

$$J(\mathbf{W}) = \|\mathbf{R} - \mathbf{U}\mathbf{W}\mathbf{\Lambda}^{1/2}\mathbf{V}^T\|_F^2 + \rho\|vec(diag(\mathbf{W}\mathbf{\Lambda}^{1/2}))\|_{L_1}, \qquad (4.3)$$

or

$$J(\mathbf{w}) = \|\mathbf{R} - \mathbf{U}\mathbf{W}\mathbf{\Lambda}^{1/2}\mathbf{V}^T\|_F^2 + \rho\|vec(\mathbf{\Lambda}^{1/2})\mathbf{W}\|_{L_1}, \qquad (4.4)$$

where $\|.\|_{l_1}$ denotes $L_1 - norm$, $vec(\mathbf{B})$ assigns the diagonal elements of $l \times d$ $\mathbf{B}$ to a vector of size $1 \times d$, and ρ is the penalty term which is a fixed Lagrange multiplier. The new update equation for estimating $\mathbf{W}$ can therefore be achieved as:

$$\mathbf{W}_{k+1} = \mathbf{W}_k - \mu \mathbf{U}\mathbf{\Lambda}^{1/2}\mathbf{V}^T(\mathbf{R} - \mathbf{U}\mathbf{W}_k\mathbf{\Lambda}^{1/2}\mathbf{V}^T)^T$$

$$+\rho'.sign[\mathbf{W}_k vec(\mathbf{\Lambda}^{1/2})].\left(vec(\mathbf{\Lambda}^{1/2})\right)^T, \qquad (4.5)$$

where $(.)^T$ stands for the transpose operation, k is the iteration index, and $\rho' = \mu.\rho$. The best initialization for $\mathbf{W}$ is the identity matrix $\mathbf{I}$, since $\mathbf{W}$ is expected ultimately to be a diagonal matrix. This is computed for each block of data separately. However, the recursive algorithm in [6] can also be used for real time processing of the data.

4.5   Comparing Basic ALE and SSA-Based ALE

To compare the new ALE with the traditional ALE, a number of experiments are carried out using both synthetic and real signals. For the synthetic data a noisy sinusoid is generated. For real data a surface electromyography (EMG) signal is used. The diagnostic information within the EMG is obscured by the effect of electrocardiogram (ECG) signal. Therefore, the problem here can also be considered as recovering of the ECG from strong noise-like EMG background. The results are presented for each experiment.

4.5.1   Simulated Signals

A four second $N = 2000$ sample sinusoid of 500Hz and 0.1 volt amplitude is generated. In one experiment Gaussian noise and in another one impulsive noise are added. The input SNR is varied between -10 to $+10$ dB and both the traditional ALE and the new proposed SSA-based ALE have been applied to the signals. Figure 4.4 presents the performance of the two ALEs in terms of output SNR with respect to input SNR for both Gaussian and impulsive noises. The output SNR is defined as:

$$SNR_{out} = \frac{\sum\limits_{t=1}^{N} \hat{y}_t^2}{\sum\limits_{t=1}^{N}(y_t - \hat{y}_t)^2}, \qquad (4.6)$$

where $\hat{y}_t$ is the estimated desired signal. It is seen that, given a narrowband signal (sinusoid) for Gaussian noise, the performances of both methods are much closer. However, when the noise is not Gaussian (impulsive) the performance of the SSA-based ALE is considerably better than that of the ALE. Figure 4.5 shows the outputs for an input $SNR = -5$ dB. The results are the average results over 10 segments of similar recordings from five subjects.

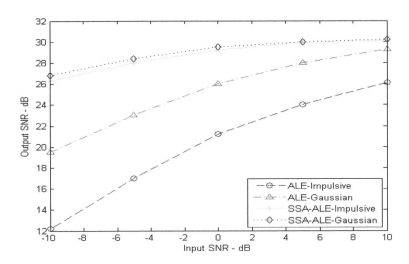

FIGURE 4.4 (See color insert.): Output SNRs for different input SNRs for synthetic data for both traditional ALE and SSA-based ALE methods.

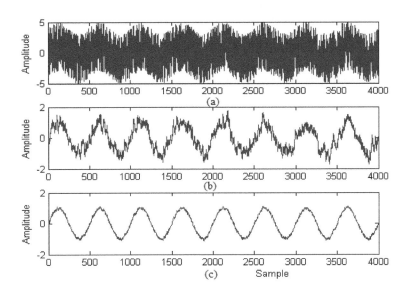

FIGURE 4.5: The outputs of the traditional and proposed ALE systems; (a) the synthetic noisy signal with an impulsive noise of 5dB SNR, (b) output of traditional ALE, and (c) output of the proposed SSA-based ALE.

4.5.2   Real Signals (EMG Corrupted by ECG Artefact)

EMG signals provide valuable information relating to peripheral and central motor functions and have been widely adopted in the study of motor function and movement disorders. Surface EMGs represent a superposition of electrical activities from motor unit action potentials located subcutaneously to the detecting electrodes. Surface EMGs have previously been applied to assess muscular activity quantitatively and diagnose various muscular disorders. These signals however, are nonstationary and often available using only single-channel recordings. Most of EMG signals are corrupted by ECG mainly because blood vessels are spread all over the body. Restoration of EMG signals from artefacts is extremely difficult, mainly due to their randomness and being noisy in nature. There have been some attempts in denoising the EMG signals recently such as that in [7] where wavelet and ICA have been used for a multichannel recording of EMG. Unfortunately, having more than one channel recording is not always possible. Moreover, generally the conditions of independence and stationarity of the signal components are not fully satisfied. In a very recent attempt by Lu et al.[9] an adaptive noise canceller (ANC) based on recursive-least-squares (RLS) algorithm was developed for removing an ECG artefact from the surface EMGs recorded in patients with cervical dystonia. Even in this approach an ECG signal recorded from a separate channel was used as a reference signal. However, so far, to the authors' knowledge, no prior attempts have been made to separate noisy ECG and EMG signals from a single channel recording without a reference component. Visual inspection of the background EMG and ECG patterns show that the ECG is nearly periodic (more periodic in shorter intervals), whereas EMG is noisy, with an unknown distribution.

An exhaustive search may be carried out over the eigenvalues. Finally select a small number of eigentriples which belong to the ECG signal and reconstruct the ECG from the selected eigenvalues. In such cases, some criteria have to be established for selection of the eigentriples of the desired signals. However, in many cases such as when the EMG is stronger than the ECG, this method won't be accurate and it is computationally demanding. One of the unsolved problems in this approach is the selection of the number of eigentriples (window size) which can vary from two to the overall number of eigentriples.

On the other hand, the use of ALE is not fully justified since the noise (EMG) distribution is not known and it is often heavy tailed. Therefore, for the above reasons, adaptive selection of the SSA eigentriples, followed in this work, is the most logical approach.

Following this approach, the ECG is extracted first. Then, the EMG component is easily obtained by subtracting the estimated ECG from the original recording;

The EMG signals were recorded from the human forearm using surface EMG. The sampling frequency was 2KHz and the subject was relatively relaxed during slow arm movements. The overall recorded data length was 70 seconds (140K data samples), divided into 20K sample segments (10 sec. each). In these experiments no preprocessing was performed. Therefore, the noisy EMG remained un-altered and the ECG was completely separated.

The period of ECG, heart rate, can be measured manually. Other techniques, such as analysing the autocorrelation function (ACF) can also be used. ACF is often used for detecting the pitch period in speech signals.

Fortunately, in this new application of SSA there is no need for selection of the number of desired eigentriples (window length) since the eigentriples are automatically weighted (selected) by the estimated $\mathbf{W}$. The only setting here is the overall number of eigentriples which can vary over a wide range without any noticeable change in the results. In this experiment we selected this number to be $d = 240$. Both traditional ALE and the proposed SSA-based

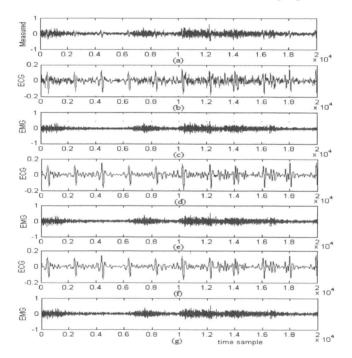

FIGURE 4.6: Separation of ECG and EMG signal in (a) using (b),(c) the traditional ALE, (d),(e) the proposed SSA-based ALE, and (f),(g) when the sparsity constraint is added to the SSA-based ALE. For each method the first signal is the separated ECG, and the second one is the EMG signal.

ALE are examined here. After processing the data using both methods the segments were joined together to reconstruct the complete ECG and EMG components. The data were obtained from 21 subjects, each recorded two times, but not necessarily from the same position, within approximately one hour time intervals. However, for illustration, only one segment of one record from one of the subjects (selected randomly with no preference) was taken. Figure 4.6 shows the results of applying traditional ALE, proposed SSA-based ALE, and finally when a sparsity constraint is added for the same signal segment. In this figure the original segment is shown in the top row. What is very important here is that the separated ECG, using the proposed methods (with or without sparsity constraint), can be directly used for clinical diagnosis. For the traditional ALE the separated ECG is not of sufficient quality to be used as an indicator of heart function.

4.6 Application of Sparsity Constraint

Often a very small number of eigenvalues is utilised for the reconstruction of cyclic data such as ECG. Therefore, further improvement is expected if the sparsity of the desired eigenvalue pattern is taken into account and used in the iterative estimation of $\mathbf{W}$. The result can be seen in the following assessment.

Since the primary objective in separation is to have independence or uncorrelatedness between the output components, a quantitative or objective assessment for comparing these two methods can be achieved. Here, the correlation coefficients of both normalized separated signals, $\zeta = \frac{1}{N} \sum_{i=1}^{N} EMG(i).ECG(i)$, $N = 140000$, are calculated. As depicted in the following table, the value of this coefficient for the traditional ALE output is $\zeta_{ALE} = 0.079$, whereas for the output of the SSA-based ALE this coefficient is negligible ($\zeta_{SSA-ALE} = 0.0016$).

TABLE 4.1: Correlation coefficient ($\pm$ max variation) between the estimated ECG and EMG for 42 data segments from 21 subjects using both traditional ALE and the SSA-based ALE methods.

Correlation Index	Traditional ALE	SSA-based ALE	SSA-based ALE+sparsity constraint
ζ	0.079 ± 0.0022	0.0018 ± 0.0003	0.0016 ± 0.0003

4.7 An Adaptive SSA-Based System for Classification of Narrow Frequency Band Signals

A supervised approach for sleep scoring from single channel EEG signals is proposed. First, a supervised SSA which is a time-frequency based method is used to extract the desired signal for each stage. Then, two recursive least squares (RLS) adaptive filters are trained and applied to EEG to identify the first and deep sleep stages.

In a recent publication, a technique able to produce the desired subspace for narrowband signals [7] has been proposed which can extract the desired component of each stage of sleep EEG signals automatically using empirical mode decomposition (EMD). This method, called SSA-EMD, exploits the information about frequency diversity of the EEG signals during various stages of sleep to select the corresponding eigen subspaces. This is due to the correspondence between particular frequency bands of the EEG signals and the sleep stages. A block diagram of this method can be seen in Figure 4.7

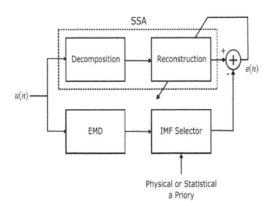

FIGURE 4.7: An EMD-SSA system for detection of signals within certain frequency bands.

Using SSA-EMD, a learning system can be developed and trained based on the EEG from different sleep stages. As each stage is characterised by a distinct set of physiological and neurological related features and a dominant frequency band, it can be considered as one good application of this method.

On the other hand, as discussed before, adaptive filters have several applications in many areas such as communications, sound, image, and signal processing [1]–[16]. They have the ability of adjusting their properties according to desired features of the input data. Therefore, they can filter the unwanted components of input data, especially in time varying signals. LMS is a very common optimization algorithm for adaptive filtering [16]. However, it has slow convergence due to the eigenvalue spread of the input data correlation matrix. On the other hand, RLS filters have a decoupled route to convergence. RLS filters can work well only if their outputs are white and Gaussian. Therefore, a hybrid multistage model has been proposed which uses RLS and SSA-EMD. This is then applied to the EEG signals for sleep scoring. At the first stage, the signal segment is passed through both RLS and SSA-EMD. The latter produces the desired, also called target, signal and the RLS updates the filter weights accordingly. To explain this in detail, first RLS recursive optimisation is reviewed here. Other adaptation techniques such as LMS, and normalised LMS (NLMS), can also be used for the same purpose.

4.7.1 Recursive Least Squares

The RLS is a recursive method that adjusts the filtering weights by minimising a weighted linear least squares cost function. If we have a desired signal $\mathbf{d}$ which is transmitted over a noisy environment, and record a noisy mixture $\mathbf{u}$ and consider the following linear equation to estimate $\mathbf{d}$ from $\mathbf{u}$ at each time instant t:

$$y(t) = \hat{d}(t) = \sum_{i=0}^{q} w_t(i)u(t-i) + v(t), \tag{4.7}$$

where $v(t)$ is the noise. RLS is used to estimate a $\mathbf{w}$ which produces an output which best resembles the target signal $\mathbf{d}$. Ignoring the noise, in vector form, the discrete domain representation is in the following form.

$$\hat{d}(n) = \mathbf{w}_n^T \mathbf{u}_n, \tag{4.8}$$

where $\mathbf{u}_n$ is a vector including p recent samples of $u(n)$. RLS has the aim of estimating the filter weights at each time sample but without performing the least square algorithm for each n. Filter coefficients, $\mathbf{w}$, for both filters are estimated using the following recursive operations [16]. Variable k in these equations refers to the kth filter, where $k = 1, 2$ in our application refers to only two bands of delta (0 to 4Hz) and alpha (8 to 12Hz) of the EEG signals respectively:

$$
\begin{aligned}
\alpha^{(k)}(n) &= d^{(k)}(n) - \mathbf{u}_n^T \mathbf{w}_{n-1}^{(k)} \\
\mathbf{g}_n^{(k)} &= \mathbf{P}_{n-1}^{(k)} \mathbf{u}_n / (\eta + \mathbf{u}_n^T \mathbf{P}_{n-1}^{(k)} \mathbf{u}_n) \\
\mathbf{P}_n^{(k)} &= \eta^{-1} \mathbf{P}_{n-1}^{(k)} \mathbf{g}_n^{(k)} \mathbf{u}_n^T \eta^{-1} \mathbf{P}_{n-1}^{(k)} \\
\mathbf{w}_n^{(k)} &= \mathbf{w}_n^{(k)} + \eta^{(k)}(n) \mathbf{g}_n^{(k)} \\
k &= 1, 2
\end{aligned}
\tag{4.9}
$$

where $0 \leqslant \eta \leqslant 1$ is the forgetting factor.

4.7.2   Application to Sleep Scoring

An adaptive system to classify the sleep stages from the scalp EEG signals can be seen in Figure 4.8. In this system, for each EEG frequency band, the signal produced based on Figure 4.7 is used as the target signal for each one of the two adaptive filters. Therefore, the aim is to train two RLS filters using SSA-EMD as a tool for estimating the desired signal at each time instant. After training, we have a set of weight vectors which can be used to classify the sleep EEG data into three stages. Therefore, the original data is separated into two parts. One part is used in the training stage, and the second part is used for testing the system. After training, the whole set of weight vectors can be used to classify the data in the test stage.

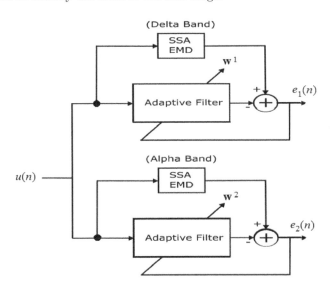

FIGURE 4.8: Two adaptive filters in parallel designed for detection of delta and alpha bands for classification of sleep stages.

4.7.3   Experimental Results

In an experimental trial thirty-six healthy men and women participated in two laboratory sessions, one involving a sleep extension protocol and the other a sleep restriction protocol. During each session polysomnography (PSG) measures were recorded at a sampling rate of 256 Hz for a baseline (BL) night (8 hours), seven condition nights (sleep extension (ES), 10 hours; sleep restriction (SR), 6 hours) and a recovery night (12 hours) following a period of total sleep deprivation. In this work the above proposed method and also one commonly used method for EEG sleep processing which uses Morlet wavelet transform, looking at time-frequency domain, were applied to sleep data to extract different frequency bands (alpha and delta; using the same technique and adding more filters the signal in other frequency bands, i.e. beta and theta can also be analysed). Delta rhythm is a slow brain wave which tends to have its highest amplitude during deep sleep in adults and is usually prominent frontally. Alpha can be seen in the posterior regions of the head on both sides and emerges with the closing of the eyes and relaxation. Therefore, only one channel for more variations in alpha and another one with more variations in delta were chosen. The signals were segmented into two parts for training and testing the system. The achieved power spectrum obtained from one segment of the test data can be seen in Figure 4.9. The results for other available signals resemble the same structure.

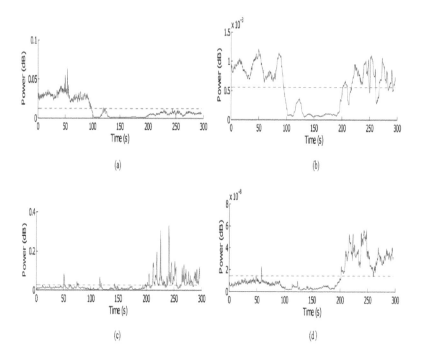

FIGURE 4.9: Two adaptive filters in parallel designed for detection of delta and alpha bands for classification of sleep stages.

From the figure, it is seen that such a method can determine the transitions between the stages of sleep by more accurately evaluating the alpha and delta (SW) brain activity variations. This achievement complies with an accurate manual scoring of the sleep data by clinical experts. For alpha rhythm, the proposed method is able to better highlight the alpha before and in the first stage of sleep and show alpha depression during deep sleep.

Further objective verification of the performance can be seen in Table 4.2. The table shows the time points where we expect to see the stage changes using manual scoring (as a bench mark) and the outcome of the methods based on EMD-SSA-Adaptive filter and wavelet transform.

TABLE 4.2:   Start of the first, the second, and slow wave stages of sleep for the examined subject obtained by manual scoring, wavelet method, and EMD-SSA-Adaptive filtering approach.

stage			
Method	First	Second	Slow wave
Manual scoring	1	101	200
Wavelet method	1	95	203
EMD-SSA-AF	1	99	200

4.8   Concluding Remarks

In the first part of this chapter a new ALE based on SSA has been introduced. It outperforms traditional ALE due to three major reasons; (1) unlike in the traditional SSA, the estimation of the filter coefficients is based on the full spectrum of SSA eigenvalues, (2) the noise does not need to be stationary or Gaussian, and (3) the periodic signal can be wideband. Moreover, there is no need for accurate setting of the SSA parameters.

Since the signal components have distinct subspaces in the eigentriple space of the augmented Hankel matrix, the effectiveness of the proposed method is considerably high.

Both ALEs have been assessed against different noise levels. The proposed system is more robust than the traditional ALE. It is also concluded that incorporating sparsity constraint leads to further but negligible improvement of the SSA-based ALE.

The proposed algorithm is applicable for the signals with some periodic components. In natural data such as biomedical signals, often the signals do not have exact duration and the periodicity can vary (often called as quasi-periodic). In order to make the algorithm applicable to such signals, two simple solutions may be proposed. The simplest one is by selection of shorter segments, within which the signals remain cyclo-stationary or periodic. A more comprehensive solution is by warping the cycle intervals around an average T for each segment of the signal. Therefore, given the cycle intervals, it is possible to warp the signals such that the cycles have the same duration. Here we assume that the source of interest has a peak in each cycle which can be detected in terms of its amplitude (either manually or automatically, using Fourier transform, periodogram, ACF, etc.). This assumption is true for some signals such as heart sound [11], or ECG signals in which the high amplitude of heart signals can be recognized easily, or even in this application which is a difficult case.

During this process firstly, the cycle durations are estimated; secondly, the average of the durations over the signal segment is calculated; thirdly, the segments with different cycle duration than the average, are warped to the average duration T, practically by applying an all-pass filter with one variable parameter.

The warping filter is often a first-order (bilinear) frequency mapping on the z-transform of the signal [12]:

$$z \longrightarrow \mathbf{A}(z) = \frac{z - \delta}{1 - \delta z}, \tag{4.10}$$

For a given sampling frequency, $-1 < \delta < 1$ is adjusted for exact mapping/warping of the signal. Dewarping the warped signal can be achieved by using the same δ of the opposite sign. Therefore, after warping the signal in each interval the value of δ can be saved for dewarping of the extracted/restored periodic signal.

This makes the proposed approach here a powerful and simple technique for the extraction of many physiological signals buried in linear, nonlinear, stationary, or nonstationary noise, or mixed with other periodic or non-periodic signals.

In the second part, a hybrid SSA-EMD system generates the desired signal for an adaptive RLS filter. Using this concept, two adaptive filters were supplied by two target signals representing delta and alpha signals extracted from two channels of EEG. This method can accurately score the EEG sleep signals in terms of the stages of sleep.

The overall chapter highlights the fact that any a priori or diversity can be nicely exploited in introducing a new adaptivity into the SSA grouping process.

Bibliography

[1] Widrow, B., Glover Jr, J. R., McCool, J. M., Kaunitz, J., Williams, C. S., Hearn, R. H., Zeidler, J. R., Dong, E. U., and Goodlin, R. C. (1975). Adaptive noise cancelling: principles and applications. *Proceedings of the IEEE*, **63**(12), 1692–1716.

[2] Golyandina, N., Nekrutkin, V., and Zhigljavsky, A. (2001). *Analysis of Time Series Structure: SSA and related techniques*. CRC Press.

[3] Alexandrov, T. H., and Golyandina, N. (2004). The automatic extraction of time series trend and periodical components with the help of the Caterpillar-SSA approach. *Exponenta Pro*, **3**(4), 54–61.

[4] GISTAT website: *www.gistatgroup.com*.

[5] Haavisto, O. (2010). Detection and analysis of oscillations in a mineral flotation circuit. *Journal of Control Engineering Practice*, **18**(1), 23–30.

[6] Ghodsi, M., Hassani, H., and Sanei, S. (2010). Extracting Fetal Heart Signal from Noisy Maternal ECG by Singular Spectrum Analysis. *Statistics and Its Interface*, **3**(3), 399–411.

[7] Teixeira, A. R., Tomé, A. M., Bohm, M., Puntonet, C. G., and Lang, E. W. (2009). How to apply nonlinear subspace techniques to univariate biomedical time series. *Instrumentation and Measurement, IEEE Transactions*, **58**(8), 2433–2443.

[8] Sanei, S., Ghodsi, M., and Hassani, H. (2011). A constrained singular spectrum analysis approach to murmur detection from heart sounds. *Elsevier Journal of Medical Engineering and Physics*, **33**(3), 362–367.

[9] Sanei, S. and Hosseini-Yazdi, A. R. (2011). Separation of single channel EMG and ECG signals using constrained singular spectrum analysis. *Proc. of the 17th Int. Conf on Digital Signal Processing (DSP)*, Corfu Greece.

[10] Ghaderi, F., Mohseni, H. R., and Sanei, S. (2011). Localizing heart sounds in respiratory signals using singular spectrum analysis. *Biomedical Engineering, IEEE Transactions*, **58**(12), 3360–3367.

[11] Azzerboni, B., Carpentieri, M., La Foresta, F., and Morabito, F. C. (2004). Neural-ICA and wavelet transform for artifacts removal in surface EMG. In *Neural Networks, 2004. Proceedings of 2004 IEEE International Joint Conference*, **4**, 3223–3228.

[12] Lu, G., Brittain, J. S., Holland, P., Yianni, J., Green, A. L., Stein, J. F., Azia, T. Z., and Wang, S. (2009). Removing ECG noise from surface EMG signals using adaptive filtering. *Neuroscience letters*, **462**(1), 14–19.

[13] Ghaderi, F., Sanei, S., Makkiabadi, B., Abolghasemi, V., and McWhirter, J. G. (2009). Heart and lung sound separation using periodic source extraction method. In *Digital Signal Processing, 2009 16th International Conference*, 1–6.

[14] Braccini, C., and Oppenheim, A. V. (1974). Unequal bandwidth spectral analysis using digital frequency warping. *Acoustics, Speech and Signal Processing, IEEE Transactions*, **22**(4), 236–244.

[15] Haykin, S. (2002). *Adaptive Filters Theory*. Wiley.

[16] Widrow, B., and Stearns, S. D. (1985). Adaptive signal processing. *Englewood Cliffs, NJ, Prentice-Hall, Inc.*, **491**, 1.

Chapter 5

Applications to Biometric Identification and Recognition

5.1   Introduction

Biometric measures are becoming prominent indicators of patient status in clinical diagnosis, human recognition, and security. Intelligent surveillance systems seek to discover possible threats automatically and raise alerts. Being able to identify the surveyed object can help determine its threat level. The current generation of devices provides digital video data to be analysed for time varying features to assist in the identification process. Commonly, people queue up to access a facility and approach a video camera in full frontal view. In this environment, a variety of biometrics are available, for example gait, which includes temporal features like stride period, possible anomalies in moving hands or walking. Gait can be measured unobtrusively at a distance. The video data will also include face features, which are short-range biometrics. In this way, one can combine biometrics naturally using one set of data.

Gaits and their recognitions have profound applications in clinical diagnosis and patient monitoring. Analysis of hand tremor for Parkinson patients or body movement for epileptic patients has become popular in clinics. In rehabilitation, treadmill stepping and its signature can provide reasonable clues on how well the patient has progressed in the treatment.

In this chapter first, a review of gait recognition and modelling with the environment is provided. Then, we discuss in detail the issues arising from deriving gait data, such as perspective and occlusion effects, together with the associated computer vision challenges of reliable tracking of human movement. Later, as the main emphasis of this chapter, application of SSA to gait recognition is covered. In the second part of this chapter, state of the art research in gait analysis for rehabilitation purposes together with application of SSA for achieving such objective is presented.

5.2   Gait Recognition

The use of biometrics has caught the imagination of the public eye. Using a measurable characteristic of a human, it has become feasible to use automated means of authentication. In a certain sense this is not new, as older analogue recognition technologies have been used for a long time. Thus, we had handwritten signatures or thumbprints. What is different now is the use of computer technology to measure or drive the features. Some examples are: face and facial features, fingerprints, hand geometry, handwriting, iris, retina, vein patterns, and voice. Another consideration is that most of the currently used biometrics require that the subject be physically close before the authentication system can work. Close-distance biometrics, for example fingerprint and iris, have high accuracy in identification rates, exceeding 99.99%. It is also becoming apparent that in large scale deployment of biometric systems, one biometric is not enough [1]. In the systems currently deployed at least two biometrics are used: fingerprint and either face or iris. This aspect will be briefly covered in a later section. A further reason for multi-biometric systems is for faster processing. In checking identities, the dataset to be checked is usually large in size. In order to reduce the computation cost, it is expedient to narrow down the space of possible identities of the subject in the recognition process. Thus, the use of other biometrics which operate at a further distance may be employed. So, for each biometric, or combination thereof, current research aims to find more robust and quicker means of obtaining the required information. In this rapidly developing field, one approach is to focus on more subtle biometrics like vein patterns, hand shape, earshapes and infra-red heat signatures. For existing biometrics, there is still much research to be undertaken on face recognition for example, using Tridimen-Sional approaches and using other features. On the other hand, concerning biometrics which can be measured from afar, gait promises to be an interesting field. With aphorisms such as "What you say reflects who you are, the way we walk is unique, etc." can be considered as an identifying trait.

Unlike several other biometrics, gait has desirable properties;

- it can operate at a large distance,

- it is often neither invasive nor intrusive, as it does not require the cooperation of the subject,

- it is hard to disguise: it has been shown that gait does not significantly vary in an individual unless there is a case of extreme physical change like carrying a load or a major change in footwear,

- different modality imaging systems may be used to record them (e.g. infrared during low light illumination).

In the context of the use of video cameras in a surveillance mode, gait is characterised for identification purposes and can be measured from a distance, far before a face can be clearly seen. Thus it can be used to pre-select a group of possible candidates to check. When the person draws near enough for face recognition to take place, the number of people to check can be drastically reduced, thus giving an overall better recognition rate. Thus a promising area in recognition is the fusing of several biometrics to make the overall process more robust. The expert knowledge embedded in each system can be brought to bear to make classification more robust. It makes sense to use as much available information as possible to perform the task of identification. A useful demarcation between the types of Gait Analyses is in order here. The study of human gait was traditionally in medical studies which were diagnostic in nature. It is a part of the field of biomechanics and kinesiology and related to other medical fields like podiatry. The gait metrics are derived from sensors attached to a human and in some cases, video sequences of gait are captured. In this respect, we may term this as Clinical Gait Analysis.

This field of study has provided much of the terminology used in gait analysis as well as the initial experiments on recognition, e.g. the various phases of a walk and the various parts of the body used in walking. In the context of the use of video cameras in a surveillance mode, gait is characterized for identification purposes. Notably, an intrusive process is not suitable. This calls for a fully automated method of tracking and analysis. For lack of a better term, we call this Video Gait Analysis and consider it as the main context of the term Gait Analysis. Of all possible approaches for human motion analysis, only a small subset has been employed for the task of human recognition by gait. This is attributable to the current limitations of technology, namely relatively low resolution video capture devices and the computationally intensive task of recovering the pose of a person given a stream of multi-dimensional images such as in video. To compound this, surveillance videos are taken at a distance so the captured images of a human are even smaller. However, with the advent of new technologies it may be possible to deploy more powerful techniques for analysis.

5.2.1   Gait Modelling and Recognition

To give some background and provide the motivation for identification by gait, we look at early experiments in this field which actually came from psychophysical studies. We start off by considering the work carried out on tracking humans and recognising walking motions. This field of research is very wide and diverse, going into the fields of virtual reality and human computer interfacing (HCI). However, since gait is a specialized human motion, features and algorithms used in motion recognition may be used to analyse gait in the same fashion.

Differentiating between various gait patterns is often difficult, as walking involves motion that is quite uniform among humans. Therefore, the objective here is to diagnose the features related to both normality and abnormality of gait. Model-free and model-based gait recognition systems are perhaps the two main categories in this field.

A comprehensive survey on human motion analysis was performed in [2]. Although dated, their taxonomy of human motion analysis has been used in many papers. A more recent survey was compiled in [3] and finally one in [4].

5.2.2   Psychophysical Considerations

Bio-mechanical and clinical studies in the 1960s show that the actions of hundreds of limbs, joints, and muscles are combined to provide an individual with the ability to walk. In addition, gait can reveal the presence of certain sicknesses, and moods and can distinguish between genders. It was shown that the variability of gait for a person was fairly consistent and not easily changed, while allowing for differentiation from others. In these early studies, there were hints about the notion of recognition by gait, but the research was mainly medically oriented. Johanssons experiments with Moving Light Displays (MLDs) [5] gave the most vivid impression of a person walking. He showed that ten lights were enough for identification of human motion. These displays comprised reflective tape attached to points of movements. The silhouette of the body was not used at all. This psychophysical study showed that a static jumble of lights makes no sense to a user, but when walking, the cadence and position of the MLDs allowed for identification of the subject as a human. For fronto-parallel (side profile) motion, the ten lights comprise: two each of elbows, wrists, knees and ankles and one each of the hip and shoulder. For more general motion, two more lights are used at the shoulder and hip level. Furthermore, Johansson showed that the bi-dimensional motion can be separated by vector analysis. Motions common to all MLDs can be decomposed into essentially translatory, rotatory, or pendulum motions. Later on, the common motions, for example, hip and shoulder motion, were either removed or modified, but still human motion was recognizable. As with earlier theories on vision, the grouping of the lights provide a gestalt effect that is learnt by human experience, namely that the total effect of the moving lights is greater than the sum of effects of the individual lights. In the same paper, he reports in his Demonstration 2, that angles from 45 to 80, can also give the impression of a walking person, implying the view-invariant nature of motion recognition.

However, whether these motions can distinguish between humans is yet to be discovered. Cutting and Kozlowski [6], in follow up experiments using MLDs, confirmed that the recognition process was direct, although observers were able to explain the features that guided their choices. Very little training was required to achieve good recognition. As compared to just looking for the presence, they show that it is possible to identify a person from the MLDs. With other collaborators, they work on attempting to identify between genders. Looking at features like shoulder/hip size ratios, torsion of the torso among others. They came up with identification of the centre of moment, a point constructed from various limb measurements. The limbs are treated as pendula attached to the torso, which is modelled as a spring. This feature accounted for 75% of the variability in gait. Finally, they postulated the sequence of events that occur when recognition due to gait takes place, starting from the centre of moment of the subject [7]. Finally, they developed a computer program to synthesise gait using the centre of moment. The graphical output was able to simulate the results obtained so far with human subjects [8]. Yu et al. [9] conducted psychology experiments on gait-based gender classification, and proposed an automated approach based on the weighted sum of block features corresponding to five parts of a body silhouette: head, chest, back, hip and leg. Little and Boyd [10] conjecture that since the frequency of all the moving points of a human are the same, it does not matter where the MLDs are. They conclude that humans do not derive structure information when viewing MLDs. However, tests have shown that when the MLD is inverted, it is not perceived as a human walking.

The question is: what mechanism does the human mind use for this task? Does it first reconstruct a tri-dimensional picture from which identification takes place? If so, this is essentially a Shape-from-Motion (SfM) problem. However, from face recognition using eigenvectors, we see that it is possible that features based on the training set itself can be used for recognition, without any prior knowledge of human features. This is the motivation for model free approaches, which use only the gross features of motion for identification purposes.

5.3   Model Free Methods

In this set of approaches to human recognition, gait can be considered to be made up of a static component based on the size and shape of a person and a dynamic component which reflects the actual movement. In gait recognition the static features are height, stride length and silhouette bounding box lengths. The dynamic features used are frequency domain parameters like frequency and phase of the movements together with parameters which identify abnormalities in walking.

The computationally expensive and detailed tasks of model reconstruction give way instead to recognizing just the gross movements of a walking person. This is normally done by obtaining the silhouette of the person to isolate the distracting background. This has been the mainstay of gait recognition approaches to date.

Two general approaches are often used; temporal comparison on a frame by frame basis and the summary of spatio-temporal information.

5.3.1   Temporal Correspondence

This is done by comparing and matching features between the test input and the existing patterns; i.e. a spatial comparison is performed temporally, on a frame by frame basis. These approaches use the image data directly, comparing the test image sequences with the training sets. In this case, the main pattern recognition approaches use correlation and hidden Markov models (HMMs).

A significant milestone in gait recognition has been the setting up of a test dataset by Sarkar et al. [11] as part of the HumanID project. A rather realistic setting was achieved by videotaping the subjects in sunlight with shadow. There were standard variations of the walker, including differences in view, type of footwear and walking surface. A manual method was used to define the bounding boxes of the walker. Then, standard algorithms for background subtracting were used to reduce noise and derive a binary silhouette of the walker with the dilation of pixels. This silhouette was then normalized in size. The foreground subtraction routine was statistically based. A baseline gait recognition algorithm was provided which allowed a probe walker sequence which was a stride cycle, to be compared to those in the gallery (database) sequences. This used a correlation between frames, using the ratio of the intersection to the union between the probe and gallery frames. A table of results shows that even a simple algorithm provided encouraging recognition rates, in the 70–80% range for simple tasks, dropping to 10–30% for difficult tasks.

Some interesting results in [11] are that the lower 20% of the silhouette, that is, from the knee down, provided for 80% of the recognition. Also, changes in surface type caused changes in walking patterns by up to five times as compared to those caused by differences in shoe type. Viewpoint did not play that much a part, presumably because the image capture was done at a distance which allowed an approximation to a fronto-parallel walk to the camera. This is considered a correlation approach.

Kale et al. [12] proposed a novel method to derive the canonical pose for walking and to apply it to a simple version of the baseline algorithm as provided in [11]. This pose is the fronto-parallel one. After showing that perspective, rather than orthographic projection is effective here, they use optical flow methods to track a feature on the walker and thereby derive the angle of walk. With an assumption of walking along a straight line, they find this angle and warp the image so it is fronto-parallel to the camera. By fusing the height information with the leg dynamics, a better recognition rate was achieved. Walking on an angle of 45 degrees still gave encouraging results.

In another work by Kale et al. [13] they used HMMs to distinguish between temporal data. They worked on a binary silhouette obtained by background subtraction and noise erosion. The feature used is the width vector, which is the difference between the left and right edges of the silhouette boundary and has as many elements as the number of rows of the image. From analysis of width vector profiles [14], only one half of a walk cycle is sufficient and provides static and dynamic information about the walker. To be of practical use, this vector needs to have about 100 elements. To compute statistics for a vector of this size, about 5,000 training samples, which is not practical, are required. To improve on this situation, several gait cycles of a walker are examined. Using k-means clustering, five images are selected from a video sequence. These images are the most representative of the stages a walker goes through in a gait cycle and are designated as stances. When a test image is obtained, a new feature vector is formed by taking the $L1$-norm of the width vector with each of the five stances, thus reducing the number of dimensions to just five. To perform identification, the sequences of images from a walker are taken and the likelihood generated from the HMMs corresponding to a particular person is computed. The highest likelihood is the most probable model then. Further experiments show that the 5-state HMM gives the best results as compared to 3 and 8-state models. This method is sensitive to changes in the viewing angle.

Ran et al. [15] presented two methods for estimating the period of gait cycles. The first is based on a Fourier transform and a periodogram, while the second uses Maximal Principal Gait Angle (MPGA) fitting. Both methods are used for pedestrian detection. Cycle characteristics are determined based on the detection of the phase difference between the input and output signals of a voltage controlled oscillator (VCO). Ho et al. [16] estimated gait cycle through both static (i.e. the highlighting of motion vector histograms) and dynamic (i.e. the extraction of Fourier descriptors) information. Dimensionality reduction of the features used was achieved through principal component analysis (PCA) and multiple discriminant analysis (MDA). A nearest neighbor classifier is then used for gait identification.

5.3.2   Spatio-Temporal Motion

In an effort to reduce the challenges of comparing images on a frame by frame basis, it is more efficient to use summarized motion features. The video signal is a bidimensional image which changes over time. A typical walking cycle may be summarized spatially, temporally or both. This can be done by obtaining the statistics of motion detected in the images in a walking cycle. In a similar way, a spatial summary may be obtained by converting the bidimensional silhouette into a single quantity which can be analysed to derive time varying features which can be used for identification. For example, Hu et al. [17] proposed the use of Gabor filters to represent body shapes through several orientations and scales. Dimensionality of the features used is reduced through the combination of PCA and Maximization of Mutual Information. Gaussian Mixture Models and HMMs taking gender into account are trained for gait classification. The authors reported the experimental results together with those of the state-of-the-art techniques. Venkat et al. [18] partitioned a silhouette into overlapping upper, middle, lower, left and right parts and used a Bayesian network for gait identification.

On the other hand, by summarizing the motion temporally as well, an image can be formed by the moving average of the value of pixels which have motion in a video frame. Thus a single motion template figure can be used for recognition using standard image comparison methods. A fixed average image may also be used. Similar motion template figures are grouped and pattern recognition is performed using methods like the k-nearest neighbor.

Lee [19] worked on the statistics of the movement of the silhouette of a person walking perpendicular to the image plane. The image of the walker is transformed into this image plane, making this a view and appearance based approach. She divided the binary level silhouette into seven elliptically shaped regions. Geometric measurements of these ellipses form the recognition features. This also introduces some noise tolerance. Using the period of variation of the features was considered. However, attempting to find the periodicity of these features was difficult at lower frame rates. Yet, features like the amplitude of leg, arm, wing and head orientation are more useful. They can be represented by the mean and standard deviation of the features.

Thus the period is not used and the final features are the mean and standard deviations of the parameters of the ellipses for a given sequence of motion. In this way, the summary of the motion is computed.

In other research, Ben-abdelkader et al. [20] tracked the subject in a video stream and extracted the bounding box of the extremes of motion. A series of such boxes is generated from the video frames at different times and normalized to a common size. These scaled boxes, known as templates, are subtracted from each other pixel by pixel (spatially) for all the given pairs of time instances (temporally). The Cutler and Davis two-dimensional similarity plot [21] is generated by plotting the sum of all the image pixel differences between frames for all time instances. This plot then is a summary of motion and is subjected to an eigenvector type of analysis for pattern recognition. Hence this was called the eigengait approach.

Wang et al. [22] considered the dynamic and static figures of a walker. The bidimensional silhouette was converted into a unidimensional distance signal based on the work of Fujiyoshi and Lipton [23]. The silhouettes were unwrapped using the geometric centroid of the silhouette as the origin of the image. A set of points along the silhouette was selected by sampling the silhouette at a fixed angular distance with respect to its origin, as shown by the lines in Figure 5.1. The sum of the Euclidean distances of these points on the silhouette from the origin form a time-varying unidimensional signal and form the bidimensional silhouette image. This distance was normalized by magnitude. A set of distance signals would thus be obtained from a walking sequence. All the distance signals from all of the walkers are subjected to eigenvector decomposition to reduce the dimension of the problem. Building on the work of Murase and Sakai [24], they also kept time stretched versions of the distance signal. A Nearest Neighbour Comparison was performed on this walker dynamic feature for recognition. Later they incorporated some static features such as height, speed, and maximum and minimum aspect ratios of the silhouettes into the recognition process. By combining static and dynamic features, they were able to obtain a 100% success recognition rate on the Soton dataset.

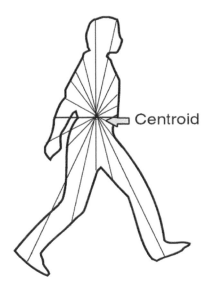

FIGURE 5.1: Unwrapping a fronto-parallel view silhouette-distance from the points on silhouette to the centroid.

In another research by Wang et al. [25] they extracted the moving walker, yielding a set of unwrapped silhouettes as described earlier in Section 5.3.1.

These silhouettes are subjected to Procrustes shape analysis. Three views have been used, from 0, 45° and 90° angles to the camera. An eigenvector analysis yielded the mean image for this cycle. This is thus a spatio-temporal summary of the walk. The mean image was used to compare the different walkers. Using a nearest neighbour approach, they reported an 88% to 90% recognition rate.

Lu et al. [26] employed three kinds of analyses on three views of unwrapped silhouettes to derive features. This is followed by independent component analysis (ICA) to reduce the data dimension. Then, they used three types of classification methods on the resulting data.

An effective approach for obtaining features from a video sequence involves summarizing the motion spatiotemporally into a motion history image (MHI) as first described by Davies and Bobick [27]. In fact, the importance of this method and its variants has been reported in [3]. The MHI is generated by identifying moving pixels in successive frames of the motion. The MHI describes how the motion proceeds by assigning a value to these pixels. In successive frames, previous moving pixels have their values decreased. Thus the most recent frames contribute brighter pixels to the MHI. The associated motion energy image (MEI) identifies where the action is taking place by performing a logical OR on the pixels between the frames. Thus, a bidimensional image now represents the motion and standard pattern recognition methods based on static images and may now be used to identify people. Based on this, Han and Bhanu [28] used gait energy images (GEI) to recognise humans by their gaits.

Zhang et al. [29] made further improvement on GEI by introducing active energy images (AEI) that concentrate on the actively moving parts of a gait silhouette, reducing the effect of movements just due to translation. They also apply two-dimensional locality preserving projections (2DLPP) to the AEI for better pattern discrimination. Silhouette images are inherently noisy and Chen et al. [30] devise the Frame Difference Energy Image to compensate for incomplete silhouettes. Missing information is derived from silhouettes in other frames which are clustered according to their GEI. The results show promise in the use of this technique.

Chen et al. [31] proposed to extract a representation based on Gabor features of a GEI on which they applied a Riemannian manifold distance-approximating projection called TRIMAP. The latter is based on the construction of a graph from data preserving pairwise geodesic distances and then optimising the discrimination ability. The reported gait recognition results outperform those of a baseline implementing latent discrimination analysis (LDA).

An interesting variant is the use of moving motion silhouette images (MMSI) by Nizami et al. [32] which summarised the motion of subsets of the silhouettes obtained for a complete gait sequence. This lead to a few MMSI for a gait sequence and these are subjected to ICA as a dimensionality reduction measure. The resulting independent components derived from each MMSI were classified using probabilistic support vector machines (SVM). The results of each SVM contribute to an accumulated posterior probability and are fused into a hybrid system to perform a more robust identification. The authors quoted a high identification rate of 100% for the NLPR (or CASIA A) dataset and 98.67% on the SotonBig dataset for their proposed methods.

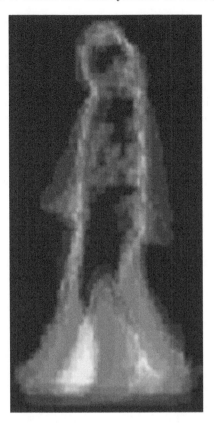

FIGURE 5.2: MEI of a person; brighter areas represent more movement.

5.3.3 Model-Based Approaches

The premise for the model based approach is the fact that interpreting the basic features in an image, such as edges and lines, require prior knowledge about the image to accelerate the process of recognition. This knowledge comes in the form of a model which can cope with ambiguities of interpretation and guide the search for movement parameters in a directed way, as we examine approaches for human recognition by gait. In research direction the waveform derived features such as frequency and amplitudes together with image features are employed, while in another approach the image pixel values are used directly.

Static and dynamic gait features may be detected, estimated, and used for gait-based human recognition. Static features represent summative variables such as mean, variance, stride length, geometry, size of body parts, silhouette geometry and amplitude of body marker variations. Dynamic features, on the other hand, demonstrate the synchrony and dynamics of movement. Long term tracking of gait dynamics may reveal the anomalies and abnormalities relating to both physical and mental diseases.

Cunado et al. [33] and Nixon et al. [34] analysed the motion of the thigh and calf, deriving the frequency and phase of the movement. By using a fast Hough transform (HT), a walker is represented by a pair of lines in a lambda shape. This affords simplification of articulated movement and takes on the HTs robustness to noise and occlusion. From the HT, the features tracked are the hip movement, modeled by a linear and oscillation term and the angle between the legs, modelled similarly by a Fourier series. Looking at the magnitude of the features, there is little variation among subjects and while phase gives better differentiation, the corresponding magnitude is small. By multiplying both the magnitude and phase, the phase-weighted magnitude is a better distinguishing feature. The authors use k-means clustering for classification with encouraging results on a small dataset.

As a later development, Foster et al. [35] attempted to introduce a degree of model specific information in the form of masks that measure the area change of a human silhouette. These masks are formed by detecting the changes in the image of a moving person and are specific to humans. By using Canonical Analysis on the AC (Alternating Current) to denote the changing, not static component of the output of the masks, they reported an 81% success rate on recognition.

Ben Abdelkader et al. [36] used stride length and cadence (walking frequency) to characterize a subject. By segmenting the walking object and constructing a bounding box around it, they were able to derive the frequency of walking by calculating the change in size of the bounding box from the video frames. The pose of the walker is found to affect the frequency of gait but could be constrained by noting the range of human gait. Later they followed this up by including height as another biometric, obtained from image measurements [37]. However, they used the fixed and variable parts of height (caused by walking) and show that the identification performance improves considerably.

Bobick and Johnson [38] looked at four static parameters of a persons walk, namely the bounding box of the walkers silhouette, the distance between the head and pelvis, the maximum distances between the pelvis and left/right foot and the distance between the right and left foot. In order to account for the variation in viewing angles, each of the parameters is scaled by a view angle dependent constant. The scale factors are measured using magnetic sensors on reference subjects in a separate experiment. The binary silhouette of the walker is derived by background subtraction. By grouping the pixels into five body regions, the effect of noise is minimized by using the centroids of these regions. Instead of using standard error measurements, the authors use the measure of expected confusion, that is how much the measurement reduces the uncertainty of identification. This measure aims to give a better idea of how the identification will perform in larger datasets. In a later development [1], the authors discussed how shadows are removed using Gaussian modelling of the image pixels, for shadow and non-shadow regions of the image. The general locations of these regions are estimated using the camera's location relative to the sun.

Tanawongsuwan and Bobick [39] investigated the use of dynamic walk characteristics (i.e. joint angle trajectories) as recognition features. They used the angles of the left and right hip and knee joints. Being an initial investigation, they used magnetic sensors attached to the subjects. The placement of such sensors could affect the results. The joint angles were calculated with care using the structure of the underlying human skeleton, so as to reduce accumulation of errors. Since they were looking at dynamic characteristics, normalizing the trajectories is needed. First, the signals are normalized so that they have unit variance. Then, they performed dynamic time warping (DTW). These signals were then time aligned to have the same duration and general shape for each trajectory of interest. The feature used is a 240 dimension vector of the four sampled trajectories concatenated. Using PCA, the size of the feature space was then reduced to 4.

In the temporal representation approaches, time is assumed as the third dimension complementing the XY axes in the image plane. A gait sequence is therefore represented in a space-time XYT three-dimensional space.

Niyogi and Adelson [40] used an XYT approach where the images from a walking sequence (XY) were stacked up to give an image cube where the third dimension was time. By analysing the XT plane, the walker's ankle traces out a unique braided pattern as compared to the head, which traces out a line. This is used to detect the presence of a walker in the image sequence. The width of the braided pattern was determined by the edges of the body and limbs as they move. The edges of the braid were found by using active contour models or snakes. The snakes were tuned to account for discontinuities at the hips, knees and ankles. The contours were formed at various levels of the XT slices giving a silhouette of the walker (in the XY plane). By averaging the silhouette, they created a stick figure representation of the walker. For recognition, they used four joint angles from the stick model. The temporal values were warped to give a common time base. They then used the derivatives of these angles and interpolated them to form a 40 dimension vector. To compare the input image sequences with that in a dataset, an $L2$-norm was used between all pairs of these sequences. They reported a 79% recognition rate.

In a later work [41], they used the XYT data directly, by fitting a so-called spatiotemporal surface to it. This surface in turn was derived from six parameters, namely the start and end positions of the walker, bounding box size of the image, period of walking and phase. This specifies the canonical surface. By analysing the actual image data, the deviation from the canonical surface is noted. Applying force causes the canonical surface to be deformed according to the subjects image data. The derived parameters describing the canonical and deviation surface provide enough information for gait classification. This approach does not face the problem of occlusion.

Kellokumpu et al. [42] made use of a three-dimensional binary representation in a spatiotemporal space to characterise the movements during walking. By extracting XYT histograms they showed improved results in comparison to state-of-the-art multi-resolution analysis of gait recognition using the CMU Mobo database. A gait sequence can be decomposed into XT slices to generate a periodic pattern termed the Double Helical Signature through an iterative curve embedding algorithm. This representation is used to highlight body parts in challenging environments for gait recognition, typically in the context of surveillance (i.e. cluttered scenes and load-carrying conditions).

In the modelling approaches, the authors propose to design models taking into account the temporal dimension. In [11] a dynamical model of human motion based on self-regression functions is used to model the gait cycle. The latter is divided into four approximately equal phases in [44]. The relation between horizontal and vertical accelerations as well as other indicators is derived by a rule-based approach following the Computational Theory of Perceptions. Zhang et al. [45] proposed two generative models representing the kinematics and visual appearances of gait through latent variables, namely the Kinematic and the Visual Gait Generative models.

There has been less research on model based gait recognition compared to model free approaches, this is probably due to the computationally intensive task of full scale pose recovery. However, it is useful to examine such approaches which derive gait by synthesis of human pose to consider features which may be used for recognition.

5.3.4   Approaches to Pose Recovery

Recovery of a human pose starts from analysis of two dimensional video images. In staying with this lower dimension, there is compactness in data representation and manipulation; the researcher works with the projection of an actual three dimensional object into a one dimensional sequence, which results in some loss of information.

Bidimensional modelling has been investigated by a number of researchers. Marr and Nishihara's early paper [46] on features for the two-dimensional representation of human figures has been influential in the area of gait recognition. They discussed various ways of representing a shape, from skeletons to a set of jointed cylinders. Later, Marr [47] gave further details from a psychophysical point of view, and a more detailed account of how to actually implement the ideas discussed.

Ran et al. [15] proposed a combination of the edge detection method and the Hough transform to extract the principal gait angle, defined as the angle between two legs during walking. A Bayesian classifier considers the frames of the gait sequence that are categorised according to whether the PGA is positively or negatively detected.

Jean et al. [48] analyzed the trajectories of the walker's head and feet. However, issues related to occlusion and looming are to be addressed. Another problem is caused by foot movement for which the separation between feet in each frame is computed to determine whether the front foot changes.

A significant representation of two-dimensional motion is the introduction of the scaled prismatic model (SPM) by Morris and Rehg [49]. In order to recover the model giving rise to a two-dimensional image stream, kinematic considerations define a state vector of joint angles as the desired objects to be derived. They also provided the mapping between these states and the two-dimensional image features. The SPM, being a projection of tridimensional objects, appeared as a link which can rotate and be scaled. Borrowing from robot manipulator theory, nonlinear least squares methods have been successful in this instance. The two-dimensional images have to be differentiable and observable. But in monocular image sequences, singularities occur in certain tridimensional configurations. The tridimensional recovery problem can be split into registration and reconstruction processes. By using the SPM in the registration process, these singularities can be better handled. The reconstruction of the tridimensional model from the SPM is done as a batch process.

Taycher et al. [50] used the SPM for motion recognition by recovering the pose of a body. By considering that the body is made up of a set of rigid segments, this is expressed as a graph, where there are as many nodes as there are segments and the edges have values of zero or one depending on whether they are connected. The nodes represent the pose of a joint of a limb. To find the optimal tree representation, they use the maximum likelihood tree that gives the lowest entropy using the maximum spanning tree algorithm. This gives the best estimate of the figure topology, i.e. the connectivity between the segments. An essential step is to calculate the pairwise mutual information between all body segments. This uses the marginal entropy of, and between, each node. In actual tests, segment locations are obtained from markers attached to a subject, and also a synthetic model was used.

Cham and Rehg [51] followed probabilistic methods for tracking the SPM links. Kalman filtering has traditionally been used to track the state vector. In realistic situations, background clutter, occlusions and complex movements give rise to a state space density function that is multi-modal. The authors contended that tracking using current Monte Carlo based techniques like the Condensation algorithm or equivalent particle filtering, incurs high computation costs. They then used a novel multiple hypothesis tracking (MHT) approach, focussing on the modes of the probability density function (PDF), assuming each mode represents a Gaussian PDF. In probabilistic tracking, it is necessary to update the PDF of the states. Using the Bayes' rule, the prior distributions are obtained by the Kalman filter for each mode. In the ensuing calculations, the modes are represented as Piecewise Gaussians rather than using the expectation maximisation (EM) algorithm to find a suitable overall PDF. Then likelihood is generated from the SPM using image pixel values. A two-dimensional, 19 degree-of-freedom SPM was initialized interactively and allowed to track a video sequence successfully.

As discussed by Di Franco et al. [52], the object motion captured by two-dimensional images in a video sequence can be considered to be degraded by noise, projection, and occlusion. Even though one can recover the SPM for the joints in two dimensions, recovery of the tridimensional motion is inherently ill-posed and some sort of regularisation needs to be imposed.

This may include kinematic constraints, which describe the connection between the links, limitation on the link lengths and restriction in the rotation axes, where applicable. This is still not enough, and the recovery process needs to be supplemented by joint angle constraints and dynamics of movement which favours smooth motion rather than any abrupt changes. Finally, interactive tridimensional key frames are specified, together with pose. All the image sequence two-dimensional SPM data are expressed as state vectors. Together with the motion constraints, these are all combined into a matrix state equation and solved for the maximum a posteriori estimate for the state that describes the motion fully.

Although a two dimensional representation of a walker's data is desirable due to its compactness, deriving the pose from two dimensional images is difficult because these images are of fully clothed and fleshed people. To match such images, more complex three dimensional models are needed.

Tridimensional modelling has also been researched by many workers in the field. The pioneering work in this field was Hogg's WALKER program [53] which works on a series of monocular images derived from video frames. The WALKER program attempts to fit the images to various proposed tridimensional models of a human walking. The proposed models for a current image frame depend on the parameters of the previous image and physical constraints of a body's movement. The model is described by a hierarchical ordering of the 14 cylinders that represent the body parts together with a specification of their movement constraints.

The procedure starts by differencing images to find the moving points. Bounding the moving object into a rectangle constrains the position of the model. It uses maximal edge points in each image as relatively invariant features and searches for correspondence to the model. A simple dataset showed successful tracking between the model and the image sequence. Rohr [54] also uses similar tridimensional models. From the images obtained from video sequences, a rectangular bounding box around the moving object is drawn. This box is used to estimate the subject's tridimensional position. Within the box, grey valued lines are found using an eigenvector technique. A series of models is generated and the contours arising from the model are matched with the grey valued lines using a geometric approach. The search for the initial model position and pose takes the longest time, but assuming smooth motion after that, it will need less resource to track the subject. The model parameters are updated using a Kalman filter approach.

Wachter and Nagel [55] fitted the projected model to the image by using the iterated extended Kalman filter (IEKF). There is a feature vector f which determines the fit between the model and the image. The differences between points on the model and image are used to correct f iteratively from one image frame to the next. A maximum a posteriori (MAP) estimation is performed on edge gradients, both in the model and image segments. By assuming that projected grey values of *surfaces* stay the same between image frames, region information is included in the update. The authors use ten parameters which are manually initialised. As well, the rates of change of the parameters are used in the update process.

Gavrila and Davis [56] used tapered super quadrics on a 22 degree-of-freedom model of a human. Using a front and side view they derived joint angles as features to match the model to the image. They segmented the moving image and back-projected it to search for the closest tridimensional model. The pose of the subject was determined by chamfer matching of edge contours [57]. However, many views of a tridimensional model need to be generated.

Bregler and Malik [58] found the motion parameters of ellipsoidal regions in the two-dimensional images of a walker. They modelled the images as scaled orthographic projections of a tridimensional model. This model is made up of tridimensional ellipsoidal body shapes. Using concepts from robotics, the pose and motion of the model was represented by twists rather than Euler angles. Since various parts of the body are linked this way, points on the various limbs form a kinematic chain related to the twists of the limbs with respect to each other, in the same manner as Euler angles. Thus, a two-dimensional point on a body segment can be expressed by so-called exponential maps which perform the required mapping of coordinates. Now, projecting the tridimensional ellipsoids creates two-dimensional support maps which help determine which pixels belong to a particular body segment. Each two-dimensional image sequence updates the model parameters by solving for the tridimensional equations by iteratively using the Newton–Raphson method after a suitable initialisation.

Using a 19 degree-of-freedom model, they were able to track various video sequences well. From this, some possible features for classification may be found in the motion parameters of the tridimensional body segments.

Leventon and Freeman [59] used a probabilistic approach, using the training data from a set of tridimensional motion sequences. Each of these was divided into ten frames which are supposed to be the most representative sub motions of human activity. This constrains the kind of tridimensional motions that generate a given sequence of two-dimensional images. These frames form a feature vector whose dimension was reduced to 50. The tridimensional data is made up of the coordinates of 37 body markers. By considering how the data is projected into a sequence of two-dimensional images, they formed an equation, using the Bayes' rule. This equation is a multidimensional Gaussian and is the posteriori probability that this two-dimensional sequence is generated by a certain set of tridimensional sequences. To test out the algorithm, a tridimensional stick figure was used which specifies what should be the two-dimensional equivalent image, using orthographic projection. The difference between the two was used to minimise an energy function. Manual intervention was also allowed in case the minimisation wouldn't proceed smoothly. The optimal set of values describing the tridimensional motion was taken and fitted to a cylindrical model for final verification of the model.

In a later development, Howe et al. [60] refined the above model in the modeling of the a posteriori equation by using a sum of Gaussians as the a priori probability distribution. To do this, the groups of tridimensional frames used as training data was grouped using k-means clustering and a Gaussian PDF was formed around each cluster. The EM algorithm was used within a MAP estimation process to find the optimal mix of Gaussians as well as the best subset of tridimensional motions describing the current two-dimensional motions. Notably, a complete sequence of two-dimensional images is a combination of subset sequences of tridimensional images. For sequences parallel to the projection direction, there is some problem with reconstructing tridimensional data, and heuristics are brought in to help. However, poor lighting, occlusion and lighting changes cause the tracker to fail in certain long sequences.

Ning et al. [61] used a tridimensional articulated body model and the joint angles and velocity as features. They used the Condensation algorithm to track these features on images parallel to the image plane only. The model had 12 degrees-of-freedom and was made up of tapered cones with a sphere. The two-dimensional images were matched based on their edge and region information [62]. In keeping with statistical methods, the a priori distribution of the features was determined beforehand by analysing the existing image sequences of walking subjects. Since they were not interested in individual walking sequences, the mean and variance of these features were obtained from all the training samples. From there they derived the class conditional distributions of the features. This served to constrain the values of the features, as we know that joint angles have limits to their values. They initialized the tracking by observing the first few frames of the test sequence, isolating the moving portion and constructing an edge map. They then correlated it with the sequences in the reference dataset in order to locate the subject. A pose evaluation function, as a measure of the posture of the subject at a given time, was computed. This was carried out by computing the degree of mismatch between the edges and regions of the model with the images. They reported good tracking, but the calculation time runs into minutes, for each frame.

Moeslund and Granum [63] implemented another Analysis-by-Synthesis approach, employing phase space to describe the motion of the model. As usual, this space is reduced by kinematic and geometric constraints corresponding to the body parts' movement and placement. The dimensionality may be reduced even more by explicitly considering the physical structure of the limbs. For example, the representation of the shoulder-elbow-hand joint can be modelled by a two-dimensional structure instead of six. The silhouette of the image is used to further reduce the search for the tridimensional model. By using a stick figure with a logical AND operation between this and the two-dimensional images, implausible poses may be further eliminated. Finally by comparing the bounding boxes of the projected tridimensional model and the two-dimensional images, further pose elimination was performed. All this allowed a reduction in dimension. The authors reported good tracking results.

From a sequence of three-dimensional gait data, Gu et al. [64] presented a method to extract the key points and pose parameters automatically. They then estimated the multiple combinations of joints and movement features as height, position and orientation of the body. HMMs were employed to model both movement and configuration features. Their goal was first to classify the actions based on a hierarchical classifier with sum and maximum a posteriori rules. Identities could then be recognised from gait sequences.

Although most gait analyses have been performed on a fronto-parallel (FP) plane to a camera, few have treated the fronto-normal (FN) plane specifically. Gait movement in the fronto-normal plane is treated as a special case of general gait movement and caters to it by the appropriate tridimensional to two-dimensional image transforms. This is done in order to take advantage of being able to observe gait from a distance of 10 m or more.

As far as experimental considerations are concerned, gait recognition is a relatively new field. Reports of successes have been mainly on small datasets and in ideal circumstances where only a single subject is present. Only recently has there been a standardised test procedure [65] on more realistic scenarios.

5.4   Combination of Gait with Other Biometrics

In the fusion of biometrics, Jain et al. [66] summarise the state of the art by describing several combinations e.g. face and speech or face and iris. The combination of biometrics takes place at several levels. In hierarchical systems, less discriminating biometrics can prune off unlikely candidates from a dataset. We allow a higher rate of false positives to be passed to other biometrics and use features that do not require many resources to calculate. This leads to a reduction in search space and thus gives a faster speed in matching. Of course if one considers the rate of success in classification, it will go up as the system works on a more likely set of candidates, and thereby compensates for the loss in accuracy, if any. In holistic systems all available data is used and combined in various ways with various rules to give an overall result [68].

Gait is a biometric that can be measured from a distance, far before a face can be clearly seen. Thus it can be used to pre-select a group of possible candidates to check. When the person draws near enough for face or even iris recognition to take place, the number of people to check can be drastically reduced, thus giving an overall better recognition rate.

Thus a promising area in recognition is the fusing of several biometrics to make the overall process more robust. It should be noted that gait seems to work best in the fronto-parallel plane. In conjunction with face features and the iris, these work with the subject facing the camera, which would seem to necessitate the use of two cameras. It is notable that the combination of face and gait biometrics is of sufficient interest that it has warranted the setting up of specific datasets. This may be due to the non-invasive nature of the images captured, and that from a distance as well. The use of face as a biometric was established by Turk and Pentland in their seminal paper [68] and this field of research is still very active. This has led to several recognition systems in use, incorporating face as a biometric. Face recognition (FR) works with images in a bidimensional realm.With the availability of hardware that makes 3D capture feasible then, paralleling the approaches in gait research, there is now work in FR in a tridimensional plane [69, 70].

Liu and Sarkar [71] use EBGM for face recognition and stance based gait recognition. The stances are derived from the silhouettes of a walker. They use Gaussian z-normalisation scores separately from face and gait recognition. They then use a variety of fusion measures like sum, Bayesian, weighted score and ranking of the sums from the two isometrics. In all cases, fusion had a beneficial effect. Most of these methods require the use of two cameras. However, some novel monocular approaches are those by Zhou and Bhanu [72] who use a profile view of a face with a gait in order to use one camera at 3.3m from the subject. For even more biometrics, the work by Bazin [73] includes the ear and footfall as well.

5.5   Gait Assessment

As briefly mentioned previously, the stride size (period) in a cyclo-stationary movement on one side and the anomaly in walking exhibited due to some physical or mental abnormalities on the other side can assist in both human recognition and abnormality diagnosis. This requires a strong time series analysis approach which can characterise both the rhythmic and non-stationary events within those series.

SSA can be effectively employed in the estimation and analysis of walking parameters as well as to detect possible anomalies. The anomalies can be the result of many mental and physical diseases such as stroke, aging, limping, or Parkinson's, where the patient's hands exhibit various tremors, and seizure. In the next few sections some applications are described.

5.6   Gait Analysis for Clinical Rehabilitation

With regard to patients with movement disabilities, a thorough clinical assessment of a patient requires estimation of the movement parameters, which in turn characterise the problem. The intensity and duration of therapy is therefore proportional to the severity of the disability.

Current assessments in use require therapists to subjectively grade various actions performed by patients, but this is an onerous and error prone process. A common solution is to use the sensors built into consumer devices to obtain patient metrics.

Being able to perform activities considered normal is important to someone who has suffered some loss of the use of their limbs. The most common non-trauma source of this is present in the patients who suffer from stroke. The inability to live an independent life requires constant medical attention, resources and often a caregiver. Where it is possible to rehabilitate the use of the limbs, a customised regimen of exercises needs to be tailored to the needs of the person, depending on the extent of the disability. At the same time, progress needs to be monitored in order to assess the effectiveness of the treatment. Presently, these are labour intensive tasks requiring trained therapists who record data, interpret them and keep track of what are often repetitive exercises. Compounding this is the lack of clinical skills at home, which allows only a limited transfer of the burden of care and may hamper the rehabilitation process [74]. One way to encapsulate the experience of healthcare practitioners is in the form of tests for limb function for tasks deemed essential in the activities of daily living (ADL). There is a large number of such established tests which involve the movements of a patient and their interactions with various objects, one of which is the focus of our work.

The use of sensors in consumer devices such as mobile phones and gaming consoles allows for a better user experience as the processors in these devices deduce the intention of the user by their movements. The use of embedded sensors and processors in objects–often in daily use–is referred to as an instrumented object approach.

Often the signals produced by the objects are processed in the frequency domain in order to detect the underlying processes of a movement. This is also true in the case of analysing biomedical signals, for which standard time or frequency analysis is used. Biological signals are never well behaved, leading to the search for newer types of analyses.

5.7 Assessing Limb Function for Stroke Patients

In setting up and establishing a test of limb functionality, enforcing a protocol for their administration provides an objective and quantitative measure of the efficacy of the patient's limb function. These measurements involve the information that can be used to document the progress of any therapy administered and to adjust its difficulty. Also, tests can be tailored to more exactly determine the nature of a disability, providing for a better regimen of therapy.

Given that, these tests should not require great expertise or time and effort to administer. Currently several of these tests use visual based scoring which introduces a degree of subjectivity and an inability to perceive subtle and possibly multiple motions. A solution to this is to automate and monitor the tests and/or the exercises through electronic means.

In such applications it is popular and desired to use low cost consumer devices and adapt them for medical use. Of particular interest are those designed for the gaming market. However, these adaptions are done on an ad-hoc basis; the availability and function of these devices are subject to the vagaries of market forces.

Instrumenting the objects used in established tests takes more resources, but it contributes to existing knowledge and faster acceptance of use based on previous ratification. As far we know this is a novel approach.

We seek to use tests that are widely accepted by the industry because they have been ratified through years of deployment. This provides a point of focus and discussion with therapists who would be familiar with the methodology used. One such test is the Action Research Arm Test (ARAT) formulated in [75] and further standardised in [76]. ARAT is a performance test designed to assess recovery of upper limb function after damage to the cerebral cortex. It can be used to check on progress in treatment as well as to evaluate the effectiveness of treatment. Its administration does not require formal training and it uses simple, short motions which can be scored and completed quickly. The test consists of various objects to be moved in a specified manner, on a table with a raised platform, with the patient seated on a chair. There are other actions as well, that do not involve objects. An assessor will score the quality of the movements, and a total score based on 19 tests provides an overall measure. A useful biomechanical aspect of the ARAT is the hierarchical nature of its assessments. This means that the movements being tested are ranked in order of difficulty in execution. In this instance, they are the Grasp, Grip, Pinch and Gross Movements which involve the fingers.

In trying to capture such fine movement, it is difficult to use methods which measure signals from the sensors attached to the subject as these may impede motion, or in the case of video give inherently noisy readings and is susceptible to the vagaries of lighting and occlusion effects.

In a work which focusses on the ARAT, it is pertinent to note that it is mainly used for assessing movement disorders caused by stroke. Of the range of movement disorders classified [77], two have been looked at. First are involuntary periodic-like muscle movements which come under several categories depending on the intensity of motion, its rapidity and the underlying causes. In this work the generic term tremor, which is well understood although the word is more commonly associated with neurological disorders, is used.

Second are distorted static hand postures which may be a result of dystonia, which are a type of hypertonia caused by the inability to control the muscle tone. Then, there are various degrees and combinations of these movements. Gross tremor frequencies occurring in the movements of the hand were around 1-4Hz and 6-11Hz. However, if the hand was weighed down, these frequencies would be reduced [78]. A higher band of frequencies at 15-30Hz were attributed to finger tremor.

Parnandi [79] describes the automated assessment of stroke patients using the Wolf Motor Function Test (WMFT) [80] detailing how the precision, smoothness and speed of movements may be scored. He uses Inertial Measurement Units (IMU) and motion capture (mocap) cameras to do this. Since the application of WMFT requires the patient to handle objects, it requires sensors to be placed on the human body for measurements. Srinivasan et al. [81] discuss the various ways sensors such as accelerometers embedded into objects used in ADL can help in tele-rehabilitation.

Lee et al. [82] built an earlier prototype of the instrumented device described in this paper which incorporates accelerometers. Portions of their paper have been reproduced here for the sake of continuity in discussion. From some recent literature, we see that obtaining live data concerning the medical condition of a person and their online analysis is a current area of research.

Blind decomposition of biomedical signals using fixed kernel methods such as by applying Fourier-based or wavelet-based decomposition does not necessary produce meaningful components.

Since various forms of movement disorders are classified based on their intensity and rapidity as mentioned earlier [77], an erroneous diagnosis may result from a frequency that is not physically present! This is a problem with decomposition methods based on basis functions decided *a priori*.

To overcome this, current frequency analyses using data driven decomposition processes using SSA has been used to analyse naturally occurring physical phenomena such as biological signals. The form of the constituent signals it produces is not constrained, for example sinusoids. SSA produces readily interpretable constituent signals from short noisy signals.

The data of the ARAT was captured using a variety of sensors. The related clinical test involved the grasping of a wooden block of $10 \times 10 \times 10$ cm^3 in size and moving it from a specified point directly to a target. The ARAT specifies this as item 1 and here it is referred to as the Cube.

This includes a sensor built into the backrest of a chair and this new set used for auxiliary measurements. Targets were placed over the sensors in the table assembly. The horizontal distance from the lower sensors to those on the platform was $0.5m$ and the sensor on the shelf was $0.25m$ from the table as seen in Figure 5.3. The sensors were connected as described in the Intersense user manual [83]. Now, two microchip microcontrollers were used as analogues to digital converters as well as data concentrators to transfer data from the two sets of sensors to a workstation via a serial link. The free scale $MMA7260$ 3-axis accelerometer was used for acceleration measurements in the Cube only and set to a $1.5g$ range with a nominal $0.8V/g$ sensitivity [85].

The two sets of serial data were streamed into a DTech $DT5070$ four serial port to USB concentrator which allowed a workstation to receive data over virtual serial ports. The sensor readings were taken at a rate of 30 samples/sec

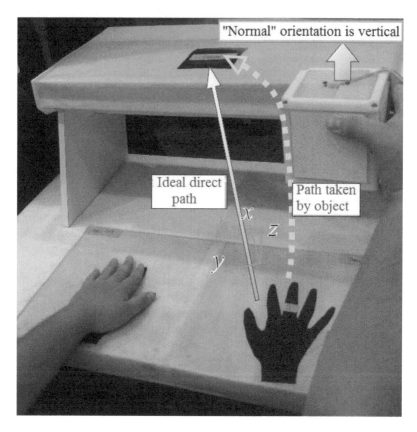

FIGURE 5.3: Cube oriented in the normal position. The ideal path of the object, compared to the actual path taken is shown. Left hand is placed on the table sensor and used as support. Two sets of force sensors are used, one in the Cube, and another in the table assembly. The image has been taken with permission from [85].

so that a frequency of up to 15Hz could be reliably recorded. As discussed before, this was sufficient for capturing the hand tremor. They used a wired connection for this round of experiments. A block diagram of the entire system is shown in Figure 5.4 [85].

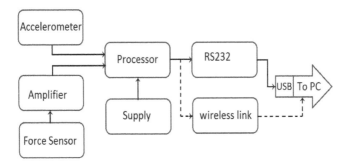

FIGURE 5.4: Hardware block diagram of Cube and table/chair embedded sensor system. Dotted lines indicate optional portions [85].

To test the system a number of healthy subjects were employed to simulate and characterise the normal and abnormal movement disorders caused by stroke, i.e. tremor and dystonia. Five healthy subjects were used and their movements analysed. The four sets of movements are repeated for five times for each person, giving a total of 100 sets of data. The test subjects were briefed as to what constitutes dystonia and tremor and to reproduce them to the best of their ability. Dystonia was simulated by not attempting to keep the Cube upright during movements. Tremors were simulated by stiffening the forearm muscles and attempting to shake the hand, which caused involuntary movement in the hands.

Later, a number of patients were examined and a similar set of data was collected from each. These patients have had a history of stroke and undergone rehabilitation. A preliminary study was undertaken on two of them. We now briefly describe the possible motions of the Cube in the ARAT as manipulated by a person.

In Figure 5.3, for the case of a normal subject, we see the Cube being grasped by a right–handed person moving it from the lower, hand silhouette to the higher black target; the trajectory is shown by a broken line. This action has to be completed in a given time. The Cube is held upright and the motion is to be what a healthy person would exert without undue duress. We would expect this task to be completed smoothly, with a minimum of energy. Note that the non-grasping (left) hand is used as support, so the force exerted on the table can also provide useful data for assessment.

In the skewed grasp and move state, loss of muscle function prevents the Cube from being held upright. Rather than use the term tilt, the authors used the term SKEW as this denotes a sense of imbalance when executing this type of movement.

Finally, in the grasp and move with tremor and/or skew, when the muscles are struggling to keep the Cube in the air, they tense up, and voluntary control is diminished. resulting in tremor. Depending on the nature of the disorder there may also be a skewed grasp on the Cube as well. Thus, for a subject, four sets of data representing the presence or absence of tremor and/or skewed grasp may be considered.

5.8   SSA Application for Recognition and Characterisation of Stroke

The waveforms obtained for a SKEW movement can be viewed in Figure 5.5. They yield detailed insights into the assessment which often cannot be obtained in any other way. First is that a patient may drop the Cube, by releasing it before placing it on the upper table. This is shown in Figure 5.5 where we note that contact was made with the bottom face before the Cube was released. Second is that the grasped surface shows peaks when just picked up and just before releasing. The peaks show concentration on the task performed and describe what Flatt [86] refers to as the power grasp. However, once the Cube is firmly grasped, the transport phase uses the precision grasp which does not use so much force. Note that, at the start time of data capture and considering the force plot, the lines with x and . markers have values that are close to zero. These are the surfaces which the hand will grasp. The non-marker line represents the force exerted by the weight of the Cube as it rests on a surface.

Thirdly, the signal given by the bottom sensor was also very useful in acting as a cue to mark the beginning of lifting the Cube and the start of a movement, and also indicate when it ends. While this was clear cut for the healthy subjects, some of the patients had problems disengaging the Cube from their grasp at the end of the move. These extraneous signals were easily detected and removed in the course of analysis, using this cue. Fourthly, an important result was that since the ARAT is a timed test, the period when the Cube is lifted can be automatically and precisely measured, and used as part of the assessment [85].

Using a Fourier transform for analysis of such data, often the quality of the results deteriorates for large datasets, resulting in a cluttered spectrum. In resorting to using SSA on the accelerometer signals however, as shown in Figure 5.6, the reconstructed components of the first five largest eigentriples have nicely discernible waveforms in the right column, even if the original signals in the left column are somewhat noisy.

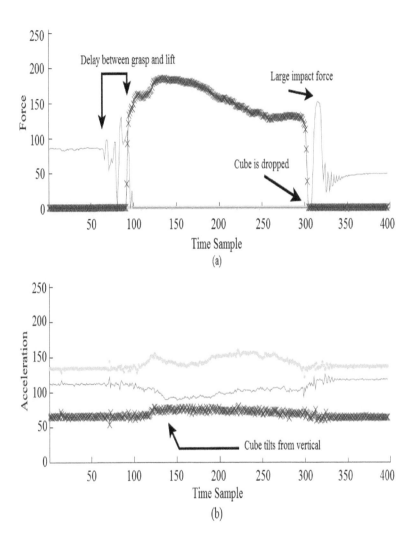

FIGURE 5.5 (See color insert.): Skew motion–note the force plot (a) where the cube is dropped rather than placed. In the (b), the tilt from the vertical is indicated [85].

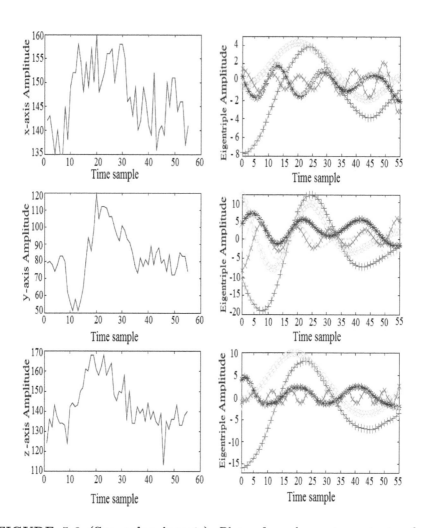

FIGURE 5.6 (See color insert.): Plots of accelerometer outputs of a healthy subject in the xyz directions in the left column. Right column has plots of the first 5 eigentriples for reconstruction of each of the accelerometer outputs [85]. The y-axis is vertical.

In SSA decomposition, the noisy parts of the waveforms are consigned to higher eigentriples which are not shown here. Next, in Figure 5.7 we can see a comparison plot between that of a healthy patient simulating a movement with tremor and skew in the left column and that of an actual patient in the right column.

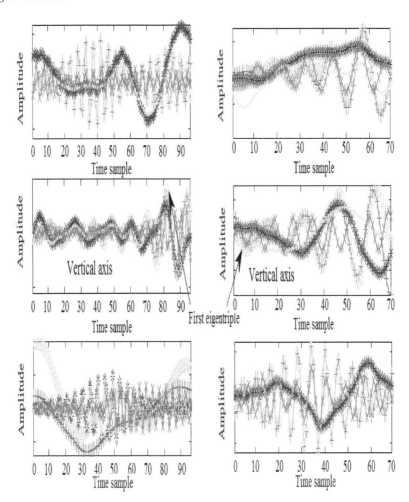

FIGURE 5.7 (See color insert.): Plots of the reconstructed waveforms for a healthy subject performing a tremor and skew movement compared with that of an actual patient in the right [85].

We observe that in the second plot which represents the Y or vertical axis, and the *second* eigentriple, marked with green o gives a good indication of the tremor movement. We denote this signal as Y_2 and note that it is a smoother waveform and thus would produce less spectral artefacts. This is an application of SSA for data smoothing. As an exploratory step, we perform an analysis on Y_2 in the following way:

- remove the signal average to have zero mean,

- transform the zero-mean signal to a frequency domain using Fourier transform,

- sort the amplitudes of the different frequency components,

- select the frequency corresponding to highest amplitude,

- for each subject, partition the data into those with and without tremor,

- remove 10% of the outliers,

- obtain the mean frequency for both movement classes.

The outlier removal step comes about because of the variability in the execution of the simulated movement. In Table 5.1, we show a portion of the results. The Trial ID is formulated as *SN-MM-T* where S is H/P for healthy subject and patient respectively, *MM* is the movement type, NM for Normal, TS for Tremor and Skew (other two not shown) and T is the trial number, 1 to 5 [85].

TABLE 5.1: The frequencies corresponding to the components with the largest amplitude for the eigentriple Y_2 of the selected trials [85].

Trial ID	Frequency of largest peak of YF_2(Hz)
H1-NM-1	1.13
H2-SK-1	0.88
H3-TR-1	6.11
H4-TS-5	6.84
P1-TS-1	2.24
p1-TS-2	1.18
P2-TS-1	1.45
P2-TS-1	1.85

In considering the entire dataset, we are able to obtain a mean frequency of 1.1 Hz with a standard deviation of 0.5 for the movements without tremor, namely the NORMAL and SKEW moves and 4.1 Hz and a deviation of 2.4 for those with tremor which are the TREMOR and TREMOR/SKEW moves. This may be interpreted that as healthy people perform normal movements, they are generally smoothly executed. However, with a simulated tremor, the execution of the motion requirement is subject to interpretation and thus can vary. In any case, with this information, we can see that except for *P1-TS-2*, stroke patients have a borderline tremor condition when moving the Cube [85].

5.9 Use of Multivariate SSA for Joint Analysis of 3D Trajectories

The above work was further improved by using multivariate SSA (MSSA). Using MSSA, it is possible to simultaneously analyse the data from all dimensions in order to identify the common harmonics, giving assurance of consistent frequency information.

For MSSA, an important difference is that the analysis is performed across *all* relevant variables. In this section we describe the process, based on [86]. For a set of D time series, at each time instant t the data is represented by a vector $x(t) = x_d(t) : d = 1 \ldots D, t = 1 \ldots N$ with N sample points [87].

Some other parameters may be estimated from the eigenvalue pattern, or so–called Scree plot. Some may involve the slope (of the regression lines) of the Scree plot at various points as another feature for patient classification [87, 88]. In these articles further improvement in stroke assessment using MSSA has been reported.

5.10 Gait Analysis Using Ear-Worn Sensor Data

A new approach to gait analysis and parameter estimation from a single miniaturised ear-worn sensor embedded with a triaxial accelerometer is introduced in [89]. In this work the subspaces of SSA are learned. SSA combined with the longest common subsequence (LCSS) algorithm has been used as a basis for gait parameter estimation. It incorporates the information from all axes of the accelerometer to estimate swing, stance and stride time parameters. Rather than only detecting local features of the raw signals, the periodicity of the signals is also taken into account. The proposed method for capturing major gait events such as the initial heel contact and toe off is validated with a high-speed camera, as well as a force-plate instrumented treadmill.

Recent advances in wearable sensing have further improved the practical use of the technique, allowing small wireless sensors to be integrated into wearable, prosthetic, and assistive devices [90, 91]. The major advantage of such sensing technology is in its ability for long-term continuous monitoring of the patient in a free-living environment, rather than specialised laboratory settings. In addition, specific context-aware gait monitoring systems can be developed to help understand the progression of a disease, assess the efficacy of the treatment and the rehabilitation process, and predict the onset of adverse events such as unstable gait patterns that may lead to high probabilities of falls in elderly patient groups.

One of the major research topics in the use of these sensors is how to balance the complexity (e.g. the number of sensors required and their practical embodiment) against the reliability and underlying information content of the platform. Naturally, the use of multiple sensors provides more information that is directly or indirectly related to the gait patterns. However, this complicates system design in terms of cross-node communication, synchronisation, and modelling. It also affects user compliance. Furthermore, consistent placement of multiple sensors is difficult, thus affecting the reliability and accuracy of the system. Such an approach, therefore, still tends to be limited to laboratory experiments.

Integrating all sensing capabilities into a single wireless sensor node has clear advantages, particularly for patient studies. Existing research has shown that the detection of certain spatio-temporal gait parameters is possible with single accelerometers. Lower trunk accelerations can be used to predict the subsequent stride's cycles and left/right steps, allowing estimation of step length and walking speed [92, 93]. Changes in gait cycle variability have been explored in musculoskeletal disorders [94]. For detailed gait analysis, other parameters such as swing and stance durations are also required. Thus far, the detection of toe off with a single accelerometer is poorly studied, and most studies are limited to multiple sensor configurations [95, 96].

With the constraint of using a single sensor, prior research has also been directed to the issue of optimal sensor placement with more interest in patient cohort comparisons. To this end, it is necessary to address the practical requirements of: 1) ease of sensor placement; 2) consistency and repeatability; 3) the underlying information content of signals. Such a problem can be treated as a feature selection problem with a multi-objective function by incorporating the above considerations and other system related constraints [97].

Previous research has shown that by placing the sensor behind the ear, most of the above constraints can be compromised or satisfied [98, 99]. It also takes advantage of the intrinsic capabilities of the skeletal bone in transmitting both high and low-frequency waves to the cranium, which can be picked up by the sensor. This, in essence, reproduces the mechanism of how humans control gait and balance. Based on this concept, we have developed an ear worn activity recognition (e-AR) sensor by using the body sensor network (BSN) platform [100]. It has been shown that gait-related force estimations, including weight acceptance and impulse, can be derived. This has been validated with a force-plate instrumented treadmill for both normal participants and patients after knee surgery [100, 101].

Thus far, the algorithms used for gait pattern analysis are limited to extraction of peaks in the raw acceleration signals, mostly to find only initial contacts. Therefore it is necessary to extract both initial contacts and toe off events from acceleration signals with accurate validation using e.g. synchronised video or force-plate related data. Therefore, it is favourable to provide a robust technique for detailed gait analysis by addressing the current drawbacks in single accelerometer-based approaches. The proposed method is model-free and based on SSA for time-series analysis [102]. It incorporates a time-series matching approach called LCSS to enhance the noise-resilience of the proposed algorithm.

The LCSS problem is to find the longest subsequence common to all of the sequences in a set of sequences. This is a computing problem [103]. The LCSS algorithm has applications in string matching and measuring similarity of two time series with different lengths [105]. LCSS is performed by the use of dynamic programming considering the matching regions in time and space. LCSS has been recursively defined in [103] as follows ; consider the two time series of **a** and **b** with variable lengths, in terms of discrete time samples, of $1 - p$ and $1 - q$ respectively,

$$
\mathbf{LCSS}_{\delta,\epsilon}(\mathbf{a}_1^p, \mathbf{b}_1^q) =
\begin{cases}
0 & \text{if } p < 1 \text{ or } q < 1 \\[2mm]
1 + \mathbf{LCSS}_{\delta,\epsilon}(\mathbf{a}_1^{p-1}, \mathbf{b}_1^{q-1}) & \text{if} \begin{cases} \|a_p - b_q\|_P < \epsilon \\ |p - q| < \delta, \end{cases} \\[4mm]
Max \begin{cases} \mathbf{LCSS}_{\delta,\epsilon}(\mathbf{a}_1^{p-1}, \mathbf{b}_1^q) \\ \mathbf{LCSS}_{\delta,\epsilon}(\mathbf{a}_1^p, \mathbf{b}_1^{q-1}) \end{cases} & \text{otherwise}
\end{cases}
$$

where $\|.\|_p$ is any p norm and δ and ε are constants set empirically. The similarity of the two time-series **a** and **b** measured by the LCSS algorithm is:

$$
S_{\delta,\epsilon}(a, b) = \frac{LCSS_{\delta,\epsilon}(a_1^p, b_1^q)}{min(p, q)},
$$

The results have shown the importance of estimated gait parameters in clinical applications such as monitoring recovery after surgery. In the following section the experimental setup and sensor placement are first described. Then, application of SSA is described. On the other hand, the LCSS algorithm is used to continue with the longest lasting subsequence.

The single-channel data is collected using a sensor containing an 8051 processor including a 2.4GHz transceiver (Nordic nRF24E1), a 3D accelerometer (Analogue Devices ADXL330), a 2MB EEPROM (Atmel AT45DB161), and a 55mAhr Li-Polymer battery [106]. The light-weight version of the e-AR that allows the recordings of mobility information and can be used in healthcare and sports applications is shown in Figure 5.8. In this figure, the axes orientation of the e-AR sensor and how it is worn by the user are also shown [102].

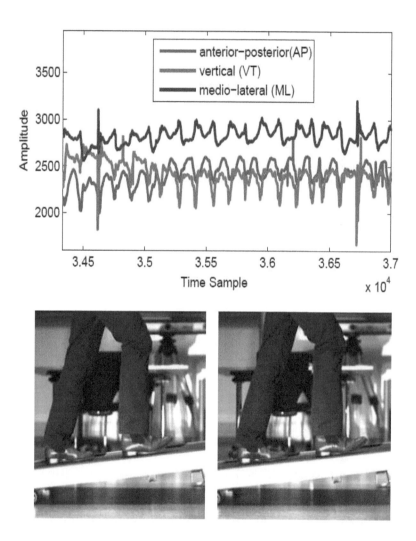

FIGURE 5.8 (See color insert.): The raw acceleration signals and the corresponding images for the large heel strikes captured by the high speed camera [102].

For some applications such as generating locomotion for paraplegic patients using functional electrical stimulation (FES), it is necessary to detect gait events in real time. In this section a general framework is proposed for real time detection of heel contacts. The proposed method can be extended to detection of both heel contact and toe-off events with proper validation to be used in related applications. In their experiment, the acceleration signals are recorded using an e-AR sensor (e-AR Lite version) with sampling frequency of 130 Hz from a healthy subject walking along the corridor both forward and returning.

The raw accelerations of three axes are shown in Figure 5.9. In Figure 5.9 a window with the size of 300 samples is used to create a subspace for the walking using SSA based on $\mathbf{U}$, $\mathbf{\Sigma}$ and $\mathbf{V}$ matrices obtained by SVD decomposition. For the next segment shown in the figure, the $\mathbf{U}$, $\mathbf{\Sigma}$, and $\mathbf{V}$ matrices are updated based on a subspace learning algorithm and the signal of the AP axis using $I_j = 1, \ldots, 30$ ($r = 30$) is reconstructed as shown in Figure 5.9 without applying SSA algorithm. The learning process has been explained in [107].

It can be seen from Figure 5.9 that subspace learning is effective for reconstruction of the acceleration signals by projecting the trajectory matrix to the learned subspace. SSA can be applied to the AP axis to find the dominant oscillation. To find the heel contacts first the signal of AP axis is selected. Then, the dominant oscillation plus the trend of the AP is formed from the first window (of size 300 samples) to create the subspace ($r = 1, 2, 3$). Then, for each new time point s_n the trajectory matrix is re-generated based on:

$$\mathbf{X}^t = [\mathbf{X}^{t-1}(x_2 : x_k)|[x_k(2)x_k(3) \ldots x_k(l)s_n]^T], \qquad (5.1)$$

where $\mathbf{X}^{t-1}, \mathbf{X}^t$ are the trajectory matrices of the previous and current iterations. A trajectory matrix considering the second and third largest eigenvalues can be generated to find the new point of the dominant oscillation based on the learning algorithm. In this step there is no need to apply the diagonal averaging since only one time point is added to the system after each iteration and the last element of the constructed trajectory matrix is extracted as the new point of the dominant oscillation, y_n. The algorithm for real-time detection of heel contacts may be summarized as follows:

1. Create the $\mathbf{U}$, $\mathbf{\Sigma}$ and $\mathbf{V}$ based on the trajectory matrix $\mathbf{X}_{I_j} = [X_1, ..., X_k]$, $I_j = 1, ..., 3$

2. For each new time point s_n do the following

    (a) update the trajectory matrix $\mathbf{X}^t_{I_j} = [\mathbf{X}^{t-1}_{I_j}(X_2 : X_k)|[X_k(2)X_k(3) \ldots X_k(l)s_n]^T]$

    (b) update $\mathbf{U}''$, $\mathbf{\Sigma}''$, and $\mathbf{V}''$ based on $\mathbf{X}^t_{I_j}$, $r = 1, \ldots, 3$, see [102]

    (c) $\mathbf{Y} = \mathbf{U}''(:, 2 : 3)\mathbf{\Sigma}''(2 : 3, :)\mathbf{V}''(k + 1 : 2k, :)^T$ as dominant oscillation $y_n = \mathbf{y}_k(l)$

    (d) Calculate $y_1 = y_n - y_{n-1}$

    (e) if $(y_1 \simeq 0)$ and $y_n < 0$, detect local minimum from multiplication of AP and SI axis using a window with the size of τ

3. update the learned subspace

Figure 5.9 compares the results of reconstruction of an AP signal using SSA and a benchmark approach called subspace projection.

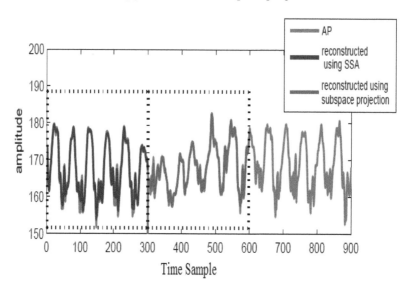

FIGURE 5.9 (See color insert.): Reconstruction of the gait signal using the real-time LCSS-SAA method [102].

From the results derived, it is evident that the method in [102] for accelerometry based gait analysis has the following advantages.

- Use of a single light-weight ear worn sensor,

- All processing can be performed on-node and therefore significantly reduces wireless-transmission overhead,

- Ergonomic design of the sensor ensures long-term patient compliance.

Another important feature of this algorithm is that for the filtering process, there are no strong assumptions on the statistics of the signals. The method can therefore be easily adapted and extended for free-living environments.

In summary, a practical gait analysis platform based on a single sensor measurement and suitable for both laboratory and free-living environments has been developed. The results derived demonstrate the practical clinical value of the method.

5.10.1 Discrimination of Gait Pattern

Jarchi and Yang [104] have analysed the acceleration signals and their characteristics for the discrimination of gait patterns. Suppose that the acceleration signals are segmented for walking down stairs, level walking and walking up stairs. Practically, using a triaxial accelerometer positioned on the ear, the important axes for analysis of different walking conditions are found to be the anterior-posterior (AP) and superior-inferior (SI) axes. By positioning the sensor on the ear, the acceleration signals are added with an artefact caused by the head movement. The signal of the AP axis for level walking for one subject is shown in Figure 5.10(a). As can be seen from this figure, the artefact caused by the head movement can be easily removed using the SSA since it appears as the trend of signal relating to the largest eigenvalue.

To start the analysis of the signals, SSA has been applied to the acceleration signal and the trend of the signal has been removed by subtracting the reconstructed trend (using the first elementary matrix) from the original signal. In this figure the raw acceleration of the AP has been shown in which the trend is shown in bold blue colour. After removing the trend of the signal and reconstructing the signal, it can be seen that the first two eigenvalues given by SSA have very close values. These eigenvalues are related to the dominant oscillation of the signal and they are essentially the second and third eigenvalues λ after applying the SSA algorithm to the raw accelerations. The dominant oscillation and raw acceleration after trend removal are shown in Figure 5.10(b,c).

After SSA and SVD, the second and third eigenvalues have been used to reconstruct the dominant oscillatory trends. Then a peak detection method is applied to estimate the local minimum points of the dominant oscillations. These signals are then segmented using the local minimum points. These cycles are grouped into a matrix.

To differentiate the level walking types (i.e. upward, and downward) from a non-level walking (i.e. horizontal), a template which is the sum of two overlapped Gaussian waveforms has been constructed [102]. This template has been formed based on prior knowledge using different types of walking and observing their effects on the acceleration signals of the AP axis. The main acceleration cycle obtained by applying SVD to the grouped cycles, has been compared to the constructed template. The comparison has been performed using the LCSS algorithm, described in the previous section. The point correspondence of the two inputs (main acceleration cycle and the template) of the LCSS given as the output of the LCSS has been used to discriminate level walking from walking down stairs and up stairs. For level walking the two points of the main cycle corresponding to the two peaks (local maxima) of the template have bigger value than the point of the main cycle corresponding to the valley (local minimum) of the template. Therefore, using the output of the LCSS, it is possible to classify and therefore detect the level walking.

In their attempt, the embedding dimension l in the SSA has been set

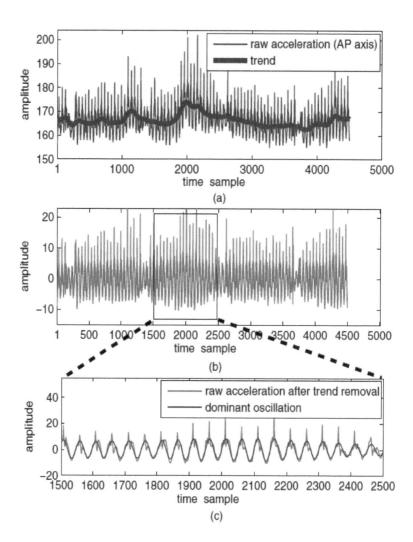

FIGURE 5.10 (See color insert.): the raw acceleration signal from the AP axis and the extracted trend by applying SSA and using the eigenvectors with the largest eigenvalue; (b) and (c) raw acceleration after trend removal and the dominant oscillation of the AP axis by using eigenvectors of the second and third largest eigenvalue respectively using raw acceleration [102].

to 100. After finding the dominant oscillation by using the local minima, the smoothed acceleration signals are segmented. To smooth the acceleration signals, SSA has been applied and the smoothed signal is reconstructed by considering $I = 2, \ldots, j = 30$ which means that the trend is removed, or in other words, the first elementary matrix relating the eigenvectors with the largest eigenvalue are ignored. Then, the cycles of the smoothed signal are grouped into a matrix and SVD is applied to extract the main cycle. In Figure 5.11 for walking downstairs, level walking and walking up stairs the smoothed acceleration cycles and the main cycles (in bold colour) are shown [102].

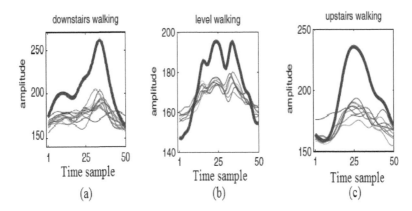

FIGURE 5.11 (See color insert.): The grouped cycles and extracted cycle as the main cycle (bold blue colour) are also shown for walking down stairs (a), level walking (b) and walking up stairs (c) [102].

A conclusion from this work is that the method is very useful for important applications such as monitoring rehabilitation of orthopaedic patients not only for detection of different walking conditions but also for analysing the effects of walking conditions on the accelerations for observing recovery over time. Time series analysis using SSA as a model free technique combined with similarity measurements using LCSS provides a strong basis for future studies of acceleration based analyses. This can include the estimation of gait parameters from raw accelerations for locating the gait events on raw accelerations [102].

5.11 Concluding Remarks

From very small number of examples here and the vast area to gait research, it is quite clear that SSA can be directly applied for gait analysis particularly for human movement characterisation, abnormality detection and gait classification. Another possible application can be the detection of a next walking step to either detect anomaly or save the patient from a danger. The advance in body sensor networking, which is likely to be integrated within fifth generation (5G) mobile communication systems, can highly benefit from such a major advantage as wearable sensor technology and the analysing algorithms involved such as SSA.

Bibliography

[1] Ross, A. A., Nandakumar, K., and Jain, A. K. (2006). *Handbook of multibiometrics*. Springer Science & Business Media, New York.

[2] Aggarwal, J. K., and Cai, Q. (1999). Human motion analysis: A review. *Computer Vision and Image Understanding*, **73**(3), 428–440.

[3] Moeslund, T. B., Hilton, A., Krüger, V. (2006). A survey of advances in vision-based human motion capture and analysis. *Computer Vision and Image Understanding*, **104**(2), 90–126.

[4] Lee, T. K., Belkhatir, M., and Sanei, S. (2014). A comprehensive review of past and present vision-based techniques for gait recognition. *Multimedia Tools and Applications*, **72**(3), 2833–2869.

[5] Johansson, G. (1973). Visual perception of biological motion and a model for its analysis. *Percept Psychophys*, **14**(2), 201–211.

[6] Cutting, J. E., Kozlowski, L. T. (1977). Recognizing friends by their walk: gait perception without familiarity cues. *Bulletin of the Psychonomic Society*, **9**(5), 353–356.

[7] Cutting, J. E., Proffitt, D. R., and Kozlowski, L. T. (1978). A biomechanical invariant for gait perception. *Journal of Experimental Psychology: Human Perception and Performance*, **4**(3), 357.

[8] Cutting, J. E. (1978). A program to generate synthetic walkers as dynamic point-light displays. *Behavior Research Methods*, **10**(1), 191–194.

[9] Yu, S., Tan, T., Huang, K., Jia, K., and Wu, X. (2009). A study on gait-based gender classification. *Image Processing, IEEE Transactions*, **18**(8), 1905–1910.

[10] Little, J. J., and Boyd, J. E. (1998). Shape of motion and the perception of human gaits. In *IEEE Workshop on Empirical Evaluation Methods in Computer Vision*.

[11] Sarkar, S., Phillips, P. J., Liu, Z., Vega, I. R., Grother, P., and Bowyer, K. W. (2005). The humanid gait challenge problem: Data sets, performance, and analysis. *Pattern Analysis and Machine Intelligence, IEEE Transactions*, **27**(2), 162–177.

[12] Kale, A., Chowdhury, A. K. R., and Chellappa, R. (2003). Towards a view invariant gait recognition algorithm. In *Advanced Video and Signal Based Surveillance, 2003. Proceedings. IEEE Conference*, 143–150.

[13] Kale, A., Rajagopalan, A. N., Cuntoor, N., and Kruger, V. (2002, May). Gait-based recognition of humans using continuous HMMs. In *Automatic Face and Gesture Recognition, 2002. Proceedings. Fifth IEEE International Conference*, 336–341.

[14] Kale, A., Sundaresan, A., Rajagopalan, A. N., Cuntoor, N. P., Roy-Chowdhury, A. K., Kruger, V., and Chellappa, R. (2004). Identification of humans using gait. *Image Processing, IEEE Transactions*, **13**(9), 1163–1173.

[15] Ran, Y., Weiss, I., Zheng, Q., and Davis, L. S. (2007). Pedestrian detection via periodic motion analysis. *International Journal of Computer Vision*, **71**(2), 143–160.

[16] Ho, M. F., Chen, K. Z., and Huang, C. L. (2009). Gait analysis for human walking paths and identities recognition. In *Multimedia and Expo, 2009. ICME 2009. IEEE International Conference*, 1054–1057.

[17] Hu, M., Wang, Y., Zhang, Z., and Wang, Y. (2010). Combining spatial and temporal information for gait based gender classification. In *Pattern Recognition (ICPR), 2010 20th International Conference*, 3679–3682.

[18] Venkat, I., and De Wilde, P. (2011). Robust gait recognition by learning and exploiting sub-gait characteristics. *International Journal of Computer Vision*, **91**(1), 7–23.

[19] Lee, L. (2002). Gait dynamics for recognition and classification. *Proc 5th IEEE Int Conf Autom Face Gesture Recognit (AFGR)*.

[20] BenAbdelkader, C., Cutler, R., Nanda, H., and Davis, L. (2001). Eigengait: Motion-based recognition of people using image self-similarity. In *Audio-and Video-Based Biometric Person Authentication*, 284–294). Springer Berlin Heidelberg.

[21] Cutler, R., and Davis, L. S. (2000). Robust real-time periodic motion detection, analysis, and applications. *Pattern Analysis and Machine Intelligence, IEEE Transactions*, **22**(8), 781–796.

[22] Wang, L., Hu, W., and Tan, T. (2002). A new attempt to gait-based human identification. In *Pattern Recognition, 2002. Proceedings. 16th International Conference*, **1**, 115-118.

[23] Fujiyoshi, H., Lipton, A. J., and Kanade, T. (2004). Real-time human motion analysis by image skeletonization. *IEICE TRANSACTIONS on Information and Systems*, **87**(1), 113–120.

[24] Murase, H., and Sakai, R. (1996). Moving object recognition in eigenspace representation: gait analysis and lip reading. *Pattern Recognition Letters*, **17**(2), 155–162.

[25] Wang, L., Ning, H., Hu, W., and Tan, T. (2002). Gait recognition based on procrustes shape analysis. In *Image Processing. Proceedings. 2002 International Conference*, **3**, 433.

[26] Lu, J., and Zhang, E. (2007). Gait recognition for human identification based on ICA and fuzzy SVM through multiple views fusion. *Pattern Recognition Letters*, **28**(16), 2401–2411.

[27] Davis, J. W., and Bobick, A. F. (1997, June). The representation and recognition of human movement using temporal templates. In *Computer Vision and Pattern Recognition, 1997. Proceedings. 1997 IEEE Computer Society Conference*, 928–934.

[28] Han, J., and Bhanu, B. (2006). Individual recognition using gait energy image. *Pattern Analysis and Machine Intelligence, IEEE Transactions*, **28**(2), 316–322.

[29] Zhang, E., Zhao, Y., and Xiong, W. (2010). Active energy image plus 2DLPP for gait recognition. *Signal Processing*, **90**(7), 2295–2302.

[30] Chen, C., Liang, J., Zhao, H., Hu, H., and Tian, J. (2009). Frame difference energy image for gait recognition with incomplete silhouettes. *Pattern Recognition Letters*, **30**(11), 977–984.

[31] Chen, C., Zhang, J., and Fleischer, R. (2010). Distance approximating dimension reduction of Riemannian manifolds. *Systems, Man, and Cybernetics, Part B: Cybernetics, IEEE Transactions*, **40**(1), 208–217.

[32] Nizami, I. F., Hong, S., Lee, H., Lee, B., and Kim, E. (2010). Automatic gait recognition based on probabilistic approach. *International Journal of Imaging Systems and Technology*, **20**(4), 400–408.

[33] Cunado, D., Nixon, M. S., and Carter, J. N. (1997). Using gait as a biometric, via phase-weighted magnitude spectra. In *Audio-and Video-based Biometric Person Authentication*, 93–102). Springer Berlin Heidelberg.

[34] Nixon, M. S., Carter, J. N., Nash, J. M., Huang, P. S., Cunado, D., and Stevenage, S. V. (1999). Automatic gait recognition. *Biometrics: personal identification in networked society*. Kluwer Academic Publishers.

[35] Foster, J. P., Nixon, M. S., and Prugel-Bennett, A. (2001). New area based metrics for gait recognition. In *Audio-and Video-Based Biometric Person Authentication*, 312–318. Springer Berlin Heidelberg.

[36] BenAbdelkader C, Cutler R, Davis L (2001) Stride and cadence as a biometric in automatic person identification and verification. *Proc Face Gesture Recognit.*

[37] BenAbdelkader, C., Cutler, R., and Davis, L. (2002). Stride and cadence as a biometric in automatic person identification and verification. In *Automatic Face and Gesture Recognition, 2002. Proceedings. Fifth IEEE International Conference*, 372–377.

[38] Bobick, A. F., and Johnson, A. Y. (2001). Gait recognition using static, activity-specific parameters. In *Computer Vision and Pattern Recognition, 2001. CVPR 2001. Proceedings of the 2001 IEEE Computer Society Conference*, **1**, 423–430.

[39] Tanawongsuwan, R., and Bobick, A. (2001). Gait recognition from time-normalized joint-angle trajectories in the walking plane. In *Computer Vision and Pattern Recognition, 2001. CVPR 2001. Proceedings of the 2001 IEEE Computer Society Conference*, **2**, 726–731.

[40] Niyogi, S. A., and Adelson, E. H. (1994). Analyzing and recognizing walking figures in XYT. In *Computer Vision and Pattern Recognition. Proceedings CVPR'94, 1994 IEEE Computer Society Conference*, 469–474.

[41] Niyogi, S. A., and Adelson, E. H. (1994). Analyzing gait with spatiotemporal surfaces. In *Motion of Non-Rigid and Articulated Objects, Proceedings of the 1994 IEEE Workshop*, 64–69.

[42] Kellokumpu, V., Zhao, G., Li, S. Z., and Pietikäinen, M. (2009). Dynamic texture based gait recognition. In *Advances in Biometrics*, 1000–1009. Springer Berlin Heidelberg.

[43] Bissacco, A., and Soatto, S. (2009). Hybrid dynamical models of human motion for the recognition of human gaits. *International journal of computer vision*, **85**(1), 101–114.

[44] Trivino, G., Alvarez-Alvarez, A., and Bailador, G. (2010). Application of the computational theory of perceptions to human gait pattern recognition. *Pattern Recognition*, **43**(7), 2572–2581.

[45] Zhang, X., and Fan, G. (2010). Dual gait generative models for human motion estimation from a single camera. *Systems, Man, and Cybernetics, Part B: Cybernetics, IEEE Transactions*, **40** (4), 1034–1049.

[46] Marr, D., and Nishihara, H. K. (1978). Representation and recognition of the spatial organization of three-dimensional shapes. *Proceedings of the Royal Society of London B: Biological Sciences*, **200**(1140), 269–294.

[47] Marr, D. (1978). Representing visual informationa computational approach. In *Computer Vision Systems*. Academic Press.

[48] Jean, F., Albu, A. B., and Bergevin, R. (2009). Towards view-invariant gait modelling: Computing view-normalized body part trajectories. *Pattern Recognition*, **42**(11), 2936–2949.

[49] Morris, D. D., and Rehg, J. M. (1998). Singularity analysis for articulated object tracking. In *Computer Vision and Pattern Recognition, 1998. Proceedings. 1998 IEEE Computer Society Conference*, 289–296.

[50] Taycher, L., Iii, J., and Darrell, T. (2002). Recovering articulated model topology from observed rigid motion. *Advances in Neural Information Processing Systems*, 1311–1318.

[51] Cham, T. J., and Rehg, J. M. (1999). A multiple hypothesis approach to figure tracking. In *Computer Vision and Pattern Recognition, IEEE Computer Society Conference*, **2**.

[52] DiFranco, D. E., Cham, T. J., and Rehg, J. M. (1999). Recovery of 3D articulated motion from 2D correspondences. *Tech. rep., Compaq Cambridge Research Lab*. Cambridge, MA, 17.

[53] Hogg, D. (1983). Model-based vision: a program to see a walking person. *Image and Vision Computing*, **1**(1), 5–20.

[54] Rohr, K. (1994). Towards model-based recognition of human movements in image sequences. *CVGIP: Image Understanding*, **59**(1), 94–115.

[55] Wachter, S., and Nagel, H. H. (1997). Tracking of persons in monocular image sequences. In *Nonrigid and Articulated Motion Workshop Proceedings*, 2–9.

[56] Gavrila, D., and Davis, L. S. (1996). Tracking of humans in action: A 3D model-based approach. In *Proc. ARPA Image Understanding Workshop*, Palm Springs, 737–746.

[57] Eigenstetter, A., Yarlagadda, P. K., and Ommer, B. (2013). Max-Margin Regularization for Reducing Accidentalness in Chamfer Matching, *Springer Lecture Notes in Computer Science*, **7724**, 152–163.

[58] Bregler, C., and Malik, J. (1998). Tracking people with twists and exponential maps. In *Computer Vision and Pattern Recognition. Proceedings. 1998 IEEE Computer Society Conference*, 8–15.

[59] Leventon, M., and Freeman, W. (1998). Bayesian estimation of 3-d human motion. *Technical Report tr 98-06*, Mitsubishi Electric Research Labs.

[60] Howe, N. R., Leventon, M. E., and Freeman, W. T. (1999). Bayesian Reconstruction of 3D Human Motion from Single-Camera Video. *NIPS*, **99**, 820–826.

[61] Ning, H., Wang, L., Hu, W., and Tan, T. (2002). Articulated model based people tracking using motion models. In *Multimodal Interface, Proceedings, Fourth IEEE International Conference*, 383–388.

[62] Ning, H., Wang, L., Hu, W., and Tan, T. (2002). Model-based tracking of human walking in monocular image sequences. In *TENCON'02. Proceedings. 2002 IEEE Region 10 Conference on Computers, Communications, Control and Power Engineering*, **1**, 537–540.

[63] Moeslund, T. B., and Granum, E. (2000). 3D human pose estimation using 2D-data and an alternative phase space representation. *Procedure Humans 2000*.

[64] Gu, J., Ding, X., Wang, S., and Wu, Y. (2010). Action and gait recognition from recovered 3-D human joints. *Systems, Man, and Cybernetics, Part B: Cybernetics, IEEE Transactions*, **40**(4), 1021–1033.

[65] Phillips, P. J., Sarkar, S., Robledo, I., Grother, P., and Bowyer, K. (2002). Baseline results for the challenge problem of HumanID using gait analysis. In *Automatic Face and Gesture Recognition, 2002. Proceedings. Fifth IEEE International Conference*, 130–135.

[66] Jain, A. K., Ross, A. A., and Nandakumar, K. (2011). *Introduction to Biometrics*. Springer Science & Business Media.

[67] Kittler, J., Hatef, M., Duin, R. P., and Matas, J. (1998). On combining classifiers. *Pattern Analysis and Machine Intelligence, IEEE Transactions*, **20**(3), 226–239.

[68] Turk, M., and Pentland, A. (1991). Eigenfaces for recognition. *Journal of Cognitive Neuroscience*, **3**(1), 71–86.

[69] Bowyer, K. W., Chang, K., and Flynn, P. (2006). A survey of approaches and challenges in 3D and multi-modal 3D plus 2D face recognition. *Computer Vision and Image Understanding*, **101** (1), 1–15.

[70] Vezzetti, E., and Marcolin, F. (2014). Geometry-based 3D face morphology analysis: soft-tissue landmark formalization. *Multimedia Tools and Applications*, **68**(3), 895–929.

[71] Liu, Z., and Sarkar, S. (2007). Outdoor recognition at a distance by fusing gait and face. *Image and Vision Computing*, **25**(6), 817–832.

[72] Zhou, X., and Bhanu, B. (2007). Integrating face and gait for human recognition at a distance in video. *Systems, Man, and Cybernetics, Part B: Cybernetics, IEEE Transactions*, **37**(5), 1119–1137.

[73] Bazin, A. I. (2006). *On probabilistic methods for object description and classification*. Doctoral Dissertation, University of Southampton.

[74] Andrews, K., and Steward, J. (1979). STROKE RECOVERY: HE CAN BUT DOES HE? *Rheumatology*, **18**(1), 43–48.

[75] Lyle, R. C. (1981). A performance test for assessment of upper limb function in physical rehabilitation treatment and research. *International Journal of Rehabilitation Research*, **4**(4), 483–492.

[76] Yozbatiran, N., Der-Yeghiaian, L., and Cramer, S. C. (2008). A standardized approach to performing the action research arm test. *Neurorehabilitation and Neural Repair*, **22**(1), 78–90.

[77] Handley, A., Medcalf, P., Hellier, K., and Dutta, D. (2009). Movement disorders after stroke. *Age and Ageing*, **38**(3), 260–266.

[78] Raethjen, J., Pawlas, F., Lindemann, M., Wenzelburger, R., and Deuschl, G. (2000). Determinants of physiologic tremor in a large normal population. *Clinical Neurophysiology*, **111**(10), 1825–1837.

[79] Parnandi, A. R. (2010). *A Framework for Automated Administration of Post Stroke Assessment Test*. MSEE thesis, University of Southern California, Los Angeles, USA, 2010.

[80] Wolf, S. L., Catlin, P. A., Ellis, M., Archer, A. L., Morgan, B., and Piacentino, A. (2001). Assessing Wolf motor function test as outcome measure for research in patients after stroke. *Stroke*, **32**(7), 1635–1639.

[81] Srinivasan, R., Chen, C., and Cook, D. (2010). Activity recognition using actigraph sensor. In *Proceedings of the Fourth International Workshop on Knowledge Discovery form Sensor Data*, Washington, DC, 25–28.

[82] Lee, T. K., Leo, K. H., and Zhang, S. (2011). Automating the assessment of rehabilitative grasp and reach. In *Information, Communications and Signal Processing (ICICS) 2011 8th International Conference*, 1–5.

[83] Birchfield, S. (1997). An elliptical head tracker. In *Signals, Systems & Amp, Computers, 1997. Conference Record of the Thirty-First Asilomar Conference*, **2**, 1710–1714.

[84] Flatt, A. E. (1983). *Care of the arthritic hand* (4th ed.). St. Louis: Moshy.

[85] Lee, T. K., Gan, S. S., Sanei, S., and Kouchaki, S. (2013). Assessing rehabilitative reach and grasp movements with Singular Spectrum Analysis. In *Signal Processing Conference (EUSIPCO), 2013 Proceedings of the 21st European*, 1–5.

[86] Flatt, A. E. (2000). Grasp. *Proc. Baylor University Medical Centre*, **13**(4), 343–348.

[87] Lee, T. K. M., Gan, S. S. W., Lim, J. G. and Sanei, S. (2014). Multivariate Singular Spectrum Analysis of Rehabilitative Accelerometry. *Proceedings of EUSIPCO 2014.*

[88] Lee, T. K. M., Gan, S. S. W., Lim, J. G. and Sanei, S. (2013). Singular Spectrum Analysis of Rehabilitative Assessment Data. *Proceedings of ICICS 2013.*

[89] Jarchi, D., Atallah, L., and Yang, G. Z. (2012). Transition detection and activity classification from wearable sensors using singular spectrum analysis. In *Wearable and Implantable Body Sensor Networks (BSN), 2012 Ninth International Conference*, 136–141.

[90] Parkka, J., Ermes, M., Korpipaa, P., Mantyjarvi, J., Peltola, J., and Korhonen, I. (2006). Activity classification using realistic data from wearable sensors. *Information Technology in Biomedicine, IEEE Transactions*, **10**(1), 119–128.

[91] Mojarradi, M., Binkley, D., Blalock, B., Andersen, R., Ulshoefer, N., Johnson, T., and Del Castillo, L. (2003). A miniaturized neuroprosthesis suitable for implantation into the brain. *Neural Systems and Rehabilitation Engineering, IEEE Transactions*, **11**(1), 38–42.

[92] Jovanov, E., Milenkovic, A., Otto, C., and De Groen, P. C. (2005). A wireless body area network of intelligent motion sensors for computer assisted physical rehabilitation. *Journal of NeuroEngineering and Rehabilitation*, **2**(1), 6.

[93] Zijlstra, W., and Hof, A. L. (2003). Assessment of spatio-temporal gait parameters from trunk accelerations during human walking. *Gait & Posture*, **18**(2), 1–10.

[94] Nilssen, R. M. and Helbostad, J. L. (2004). Estimation of gait cycle characteristics by trunk accelerometry. *Journal of Biomechanics*, **37**, 121–126.

[95] Tochigi, Y., Segal, N. A., Vaseenon, T., and Brown, T. D. (2012). Entropy analysis of tri-axial leg acceleration signal waveforms for measurement of decrease of physiological variability in human gait. *Journal of Orthopaedic Research*, **30**(6), 897–904.

[96] Willemsen, A. T. M., Bloemhof, F., and Boom, H. B. (1990). Automatic stance-swing phase detection from accelerometer data for peroneal nerve stimulation. *Biomedical Engineering, IEEE Transactions*, **37**(12), 1201–1208.

[97] Jasiewicz, J. M., Allum, J. H., Middleton, J. W., Barriskill, A., Condie, P., Purcell, B., and Li, R. C. T. (2006). Gait event detection using linear accelerometers or angular velocity transducers in able-bodied and spinal-cord injured individuals, *Gait & Posture*, **24**(4), 502–509.

[98] Selles, R. W., Formanoy, M. A., Bussmann, J., Janssens, P. J., and Stam, H. J. (2005). Automated estimation of initial and terminal contact timing using accelerometers; development and validation in transtibial amputees and controls. *Neural Systems and Rehabilitation Engineering, IEEE Transactions*, **13**(1), 81–88.

[99] Atallah, L., Lo, B., King, R., and Yang, G. Z. (2011). Sensor positioning for activity recognition using wearable accelerometers. *IEEE transactions on Biomedical Circuits and Systems*, **5**(4), 320–329.

[100] Yang, G. Z. (2006). *Body Sensor Networks*. Springer-Verlag, Berlin, Germany.

[101] Atallah, L., Wiik, A., Jones, G. G., Lo, B., Cobb, J. P., Amis, A., and Yang, G. Z. (2012). Validation of an ear-worn sensor for gait monitoring using a force-plate instrumented treadmill. *Gait & Posture*, **35**(4), 674–676.

[102] Jarchi, D., Wong, C., Kwasnicki, R., Heller, B., Tew, G., and Yang, G. (2014). Gait Parameter Estimation From a Miniaturized Ear-Worn Sensor Using Singular Spectrum Analysis and Longest Common Subsequence, *IEEE Transactions on Biomedical Engineering*, **61**(4), 1261–1273.

[103] Bergroth, L., Hakonen, H., and Raita, T. (2000). A survey of longest common subsequence algorithms. In *String Processing and Information Retrieval, 2000. SPIRE 2000. Proceedings. Seventh International Symposium*, 39–48.

[104] Jarchi, D., and Yang, G. Z. (2013). Singular spectrum analysis for gait patterns. In *Body Sensor Networks (BSN), 2013 IEEE International Conference*, 1–6.

[105] Das, G., Gunopulos, D., and Mannila, H. (1997). Finding similar time series. In *Principles of Data Mining and Knowledge Discovery*, 88–100. Springer Berlin Heidelberg.

[106] Lo, B., Pansiot, J., and Yang, G. Z. (2009). Bayesian analysis of sub-plantar ground reaction force with BSN. In *Wearable and Implantable Body Sensor Networks, 2009. BSN 2009. Sixth International Workshop*, 133–137.

[107] Golub, G. H., and Van Loan, C. F. (1996). *Matrix Computations*. The Johns Hopkins University Press.

[108] Ahad, M. A. R., Tan, J. K., Kim, H., and Ishikawa, S. (2012). Motion history image: its variants and applications. *Machine Vision and Applications*, **23**(2), 255–281.

Chapter 6

Complex-Valued SSA for Detection of Event Related Potentials from EEG

Many signals and systems are inherently complex. Radio frequency digital communication signals with multidimensional constellations are good examples. Many other signals of multiple components can also be represented in complex form to investigate their orthogonality, synchronization, coherency, circularity, etc.

There are also many ways of generating complex signals by combining real data to jointly analyse them using complex-valued mathematics. The notion of complex numbers is intimately related to the *Fundamental Theorem of Algebra* rooted in the foundation of mathematical analysis. The development of complex algebra, however, is not as simple as treating real-valued algebra for many applications.

In the case of adaptive systems the signal magnitude is often used as the main source of information. Real world processes with the 'intensity' and 'direction' components (radar, sonar, vector fields), however, also require the phase information to be considered. In the complex domain, the phase information is accounted for naturally, and this chapter illustrates the duality between processing real and complex domains for several classes of real world processes. It is shown that the advantages of using complex valued solutions for real valued problems arise not only from the full utilisation of the phase information (e.g. time-delay converted into a phase shift), but also from the use of different algebra and statistics.

A number of works in the complex-valued signal processing domain have been reported recently [1]. Good introductions to complex-valued signal processing can be found in [2] and [3]. One area where optimisation problems with complex-valued matrices appear is digital communications, in which digital filters may contain complex-valued coefficients [4]. Other areas include analysis of power networks and electric circuits [5]; control theory [6]; adaptive filters [7]; resource management [8]; sensitivity analysis [9]; and also a number of examples in acoustics, optics, mechanical vibrating systems, heat conduction, fluid flow, and electrostatics [10].

As in most of the adaptive algorithms the parameters being estimated have to be optimised. Convex optimization of complex-valued parameters has been treated in [11] and [12]. Usually, using complex-valued matrices leads to fewer computations and more compact expressions, compared with treating the real and imaginary parts as two independent real-valued matrices. The complex-valued approach is more general and usually easier to handle than working with the real and imaginary parts separately, because the complex matrix variable and its complex conjugate should be treated as independent variables when complex-valued matrix derivatives are calculated. Such an approach is important since the necessary conditions for optimality can be found through these derivatives. By setting the complex-valued matrix derivative of the objective function equal to zero, necessary conditions for optimality are found [1].

The approaches involved in *generalised* complex-valued matrix derivatives cover matrices with types of structure such as complex-valued diagonal, symmetric, skew-symmetric, Hermitian, skew-Hermitian, orthogonal, unitary, and positive semi-definite matrices. Finding derivatives with respect to complex-valued structured matrices is related to the field of manifolds. Optimization over such complex-valued constrained matrix sets can be performed by using the theory of generalised matrix derivatives [1].

6.1   Brief Literature

An early contribution to real-valued symbolic matrix calculus, found in [1], presents a basic treatment of matrix derivatives. Matrix derivatives in multivariate analysis are presented in [14] which is followed by [15] emphasising the statistical applications of matrix derivatives.

Another early thought on the analytical aspects of complex numbers was by Georg Friedrich Bernhard Riemann (1826–1866). The established principles are nowadays the basics behind what is known as manifold signal processing. To illustrate the potential of complex numbers in this context, consider the stereographic projection [32] of the Riemann sphere, shown in Figure 6.1(a).

In a way analogous to Cardano's 'depressed cubic', we can perform dimensionality reduction by embedding C in R^3, and rewriting

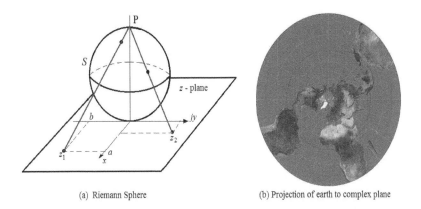

(a) Riemann Sphere   (b) Projection of earth to complex plane

FIGURE 6.1: Stereographic projection and Riemann sphere: (a) the principle of the stereographic projection; (b) stereographic projection of the Earth (seen from the south pole) [2].

$$Z = a + jb, \quad (a, b, 0) \in \mathbb{R}^3.$$

There is a one-to-one correspondence between the points of $\mathbb{C}$ and the points of S, excluding P (the north pole of S), since the line from any point $z \in \mathbb{C}$ cuts P in precisely one point. If we include the point ∞, so as to have the *extended complex plane* $\mathbb{C} \cup \infty$, then the north pole P from sphere S is also included and we have a mapping of the Riemann sphere onto the extended complex plane. A stereographic projection of the Earth onto a plane tangential to the north pole P is shown in Figure 6.1 (b).

The first fundamental work of Wirtinger [16] showed that the complex variable and its complex conjugate can be treated as independent variables when finding derivatives.

A pioneering work on using complex-valued adaptive algorithms is the one by Widrow, Mc Cool and Ball [33], as an extension of the real LMS in 1975, titled *Complex Least Mean Squares* (CLMS).

6.2 Complex-Valued Matrix Variables **Z** and **Z***

One way of representing complex-valued input matrix variables is by the use of two matrices $\mathbf{Z} \in \mathbb{C}^{N \times Q}$ and $\mathbf{Z}^* \in \mathbb{C}^{N \times Q}$. In this chapter, it is assumed that all the elements within $\mathbf{Z}$ are independent. Also the elements within

$d\mathbf{Z}$ and $d\mathbf{Z}^*$ are linearly independent. None of the matrices $d\mathbf{Z}$ and $d\mathbf{Z}^*$ is a function of $\mathbf{Z}$ and $\mathbf{Z}^*$ and, hence, their differentials are the zero matrix. Mathematically, this can be formulated as [1].

$$d^2\mathbf{Z} = d(d\mathbf{Z}) = 0_{N \times Q} = d(d\mathbf{Z}^*) = d^2\mathbf{Z}^*.$$

The representation of the input matrix variables as $\mathbf{Z}$ and $\mathbf{Z}^*$ has been exploited to develop a theory for finding Hessians of complex-valued *scalar* functions [1]. Also, an alternative representation of the complex-valued matrix variables will be presented. It will be used to simplify the process of finding complex Hessians of scalar, vector, and matrix functions.

6.3   Augmented Complex Statistics

It has been realised that the statistics in a complex domain are not an extension of the corresponding statistics in a real domain [2]. As examples, the so called 'conjugate linear' (also known as widely linear [34]) filtering was introduced by Brown and Crane in 1969 [35], generalised complex Gaussian models were introduced by Van Den Bos in 1995 [36], and the notions of proper complex random process (closely related to the notion of 'circularity') and "improper complex random process" were introduced by Neeser and Massey in 1993 [37].

Some other useful outcomes on 'augmented complex statistics' include work by Schreier and Scharf [38]–[40], and Picinbono, Chevalier and Bondon [41, 42]. This work has given rise to the application of augmented statistics in adaptive filtering, both supervised and blind. For supervised learning, extended Kalman filtering (EKF) based training in the framework of complex-valued recurrent neural networks (NNs) was introduced by Goh and Mandic in 2007 [43]. On the other hand, augmented learning algorithms within a stochastic gradient framework were proposed by the same authors in [44].

Algorithms for complex-valued blind separation problems in biomedicine were introduced by Calhoun and Adali [45, 46], whereas Eriksson and Koivunen focussed on the applications in communications [47, 48]. A statistical test for complex-valued source separation was suggested by Gautama, Mandic and Van Hulle [49], whereas a test for complex circularity was developed by Schreier, Scharf and Hanssen [50]. The recent book by Schreier and Scharf gives an overview of complex statistics [51].

In [17], a theory is developed for finding derivatives of complex-valued scalar functions with respect to complex-valued *vectors*. In this work it has been argued that it is better to use the complex-valued vector and its complex conjugate as input variables instead of the real and imaginary parts of the vector, the main reason being that the complex-valued approach often leads to a simpler approach that requires fewer calculations than the method that treats the real and imaginary parts explicitly.

An introduction to matrix derivatives, which focusses on component-wise derivatives and to the Kronecker product, is found in [18]. More on component-wise derivatives of both real-valued and complex-valued matrix derivatives can be found in [19]. Several useful results on complex-valued matrices are collected into [20], which also contains a few results on matrix calculus for which a component-wise treatment was used.

Magnus and Neudecker [21] give an insightful theory of real-valued matrices with independent components. However, they do not consider the case of formal derivatives, where the differential of the complex-valued matrix and the differential of its complex conjugate should be treated as *independent*; moreover, they do not treat the case of finding derivatives with respect to complex-valued *patterned matrices* (i.e., matrices containing certain structures). The problem of finding derivatives with respect to *real-valued* matrices containing independent elements is well known and has been studied, for example, in [22], [23], and Chapter 10 of [24]. A systematic and simple way to find derivatives with respect to *unpatterned* complex-valued *matrices* is presented in [25]. More on first and second order derivatives of (patterned and unpatterned) complex-valued matrices can be found in [26, 31].

In the following section, a brief explanation of augmented complex-valued matrix variables is provided. For further understanding the concepts the reader should refer to the references herein.

6.4 Analytic Signals

One convenient way to obtain the phase and instantaneous frequency information from a single channel recording $x(t)$ is by means of an *analytic* extension of a real valued signal. The basic idea behind analytic signals is that due to the symmetry of the spectrum, the negative frequency components of the Fourier transform of a real valued signal can be discarded without loss of information. For instance, a real valued cosine wave $x(t) = \cos(\omega t + \varphi)$, can be converted into the complex domain by adding the phase shifted 'phase-quadrature' component $y(t) = \sin(\omega t + \varphi)$, as an imaginary part, to give

$$z(t) = x(t) + jy(t) = e^{j(\omega t + \phi)}, \qquad (6.1)$$

Unlike cosine and sine which have spectral components in both positive and negative frequency ranges, their analytic counterpart $z(t) = e^{j(\omega t + \phi)}$ has only one spectral component in the positive frequency range.

An arbitrary signal can be represented as a weighted sum of orthogonal harmonic signals. Therefore, the analytic transform also applies to any general signal, however, instead of a simple phase shift of $\pi/2$ one needs to employ a filter to perform the so-called Hilbert transform to give

$$z = x + jH(x) = x + jy, \tag{6.2}$$

where $H(.)$ denotes the Hilbert transform [52]. This transform in time domain is as follows:

$$y(t) = x(t) * h(t) = \int_{-\infty}^{\infty} \frac{x(\tau)}{t - \tau} d\tau, \tag{6.3}$$

where $*$ indicates convolution. The corresponding filter can also be presented in a frequency domain as:

$$\mathbf{H}(\omega) = \begin{cases} e^{+j\pi/2} & \text{for } \omega < 0 \\ 0 & \text{for } \omega = 0 \\ e^{-j\pi/2} & \text{if } \omega > 0 \end{cases}$$

where $H(\omega)$ is Fourier transform of $h(t)$.

Although complex valued representations may not have direct physical relevance (only their real parts do), they can provide a general and mathematically more tractable framework for the analysis of several important classes of real processes [2]. Two aspects of this duality between real and complex valued processes are particularly important:

- *phase* (time delay) information (communications, array signal processing, beamforming);

- The advantages arising from the simultaneous modelling of the '*intensity*' and '*direction*' component of vector field processes in $\mathcal{C}$ (radar, sonar, vector fields, wind modelling).

Given the above properties, due to many phenomena being naturally presentable in complex domain, a large number of applications within that domain can be referred to. As an example, complex domain processing of real data has significant potential in the design of brain computer interface (BCI). In the design of brain prosthetics, a major problem in the processing information from a microarray of electrodes implanted into the cortex, is the modelling of neuronal spiking activity. High noise level often makes the analysis in the real domain rather difficult. However, the synchronisation of spike events is straightforward to model in the complex domain. An approach for converting multichannel real valued sequences of spiking neuronal recordings (point processes coming from implanted microarrays in the brain cortex) into their complexvalued counterparts is elaborated in [53]. The main underlying idea is to code the spike events of interest as complex processes, where the phase encodes the interspike interval.

Another application is in analysis of magnetic resonance imaging (MRI) which is acquired as a quadrature signal, by means of two orthogonally placed detectors and is then Fourier transformed for further processing [54]. Conventional approaches for the enhancement of MRI images consider only the magnitude spectra of such images. The phase information, however, can be obtained as a function of the difference in magnetic susceptibility between a blood vessel and the surrounding tissue as well as from the orientation of the blood vessel with respect to the static magnetic field. Recent results of the works by Calhoun and Adali [45, 46] illustrate the benefits of incorporating both the magnitude and phase information into the processing of functional MRI (fMRI) data.

In *Mathematical biosciences*, in *functional genomics*, problems such as temporal classification of confocal images of expression patterns of genes can be solved by NNs with multivalued neurons (MVN). These artificial neurons are processing elements with complex valued weights and high functionality which have proved to be efficient in image recognition problems [55]. The goal of temporal classification is to obtain groups of embryos which are indistinguishable with respect to the temporal development of expression patterns.

6.5   Complex-Valued Derivatives

One of the most important aspects in application of complex-valued computations which is often used in optimisation and adaptive processing is the calculation of derivatives. First order derivations of a complex function $f(x, y)$ with respect to a complex value z and its conjugate z^* is given as:

$$\frac{\partial f}{\partial z} = \frac{1}{2}\left(\frac{\partial f}{\partial x} - j\frac{\partial f}{\partial y}\right), \tag{6.4}$$

$$\frac{\partial f}{\partial z^*} = \frac{1}{2}\left(\frac{\partial f}{\partial x} + j\frac{\partial f}{\partial y}\right). \tag{6.5}$$

In the cases of differentiation of a scalar function f with respect to matrices, it is important to know that the two real-valued matrix variables $\mathrm{Re}(\mathbf{Z} = \mathbf{X})$ and $\mathrm{Im}(\mathbf{Z} = \mathbf{Y})$ are independent of each other, and, hence, are their differentials. Although the complex variables $\mathbf{Z}$ and $\mathbf{Z}$ are related, their differentials are linearly independent [1]. The idea of identifying the first-order complex-valued matrix derivatives from the complex differential is the key procedure for finding matrix derivatives.

If $f(z) = u(x, y) + jv(x, y)$, for $f(z)$ to be analytic in z not only u and v

should be differentiable with respect to x and y but also the so–called Cauchy–Riemann equations/conditions should be satisfied, i.e.

$$\frac{\partial u(x,y)}{\partial x} = \frac{\partial v(x,y)}{\partial y}, \quad \frac{\partial v(x,y)}{\partial x} = -\frac{\partial u(x,y)}{\partial y}. \tag{6.6}$$

This requires independency of x and y. Unfortunately, in many applications this does not happen and the Cauchy–Riemann equations are not valid and therefore, the complex functions are not differentiable. Generalised derivatives of functions of complex variables are then defined in order to proceed with analysis of such systems [2].

6.6 Generalized Complex-Valued Matrix Derivatives

Although traditionally complex-valued problem solving in signal processing relied on independency of real and imaginary components, often problems appear for which we have to find a complex-valued matrix that minimises or maximises a real-valued objective function under the constraint that the matrix belongs to a set of matrices with a structure or pattern (i.e. where there exist some functional dependencies among the matrix elements). The basic theory presented above is not suitable for the case of functional dependencies among elements of the matrix. In Chapter 6 of [1] a systematic method has been presented for finding the generalised derivative of complex-valued matrix functions, which depend on matrix arguments that have a certain structure.

Frequently encountered complex functions of complex variables which are not analytic include those which depend on the complex conjugate, and those which use absolute values of complex numbers. For example, the complex conjugate $f(z) = z^*$ is not analytic, and the Cauchy–Riemann conditions are not satisfied, as can be seen from its Jacobian

$$\mathbf{J} = \begin{bmatrix} 1 & 0 \\ 0 & -1 \end{bmatrix}. \tag{6.7}$$

This is also the case with the class of functions which depend on both $z = x + jy$ and $z^* = x - jy$. Consider for example [2]

$$J(z, z^*) = zz^* = x^2 + y^2 \;\Rightarrow\; \mathbf{J} = \begin{bmatrix} 2x & 2y \\ 0 & 0 \end{bmatrix} \;\Leftrightarrow\; \frac{\partial u}{\partial x} \neq \frac{\partial v}{\partial y}\, \frac{\partial v}{\partial x} \neq -\frac{\partial u}{\partial y} \tag{6.8}$$

This concludes that, any polynomial in both z and z^*, or any polynomial of z^* alone, is not analytic. Hence, the cost function, mean squared error, often used for optimisation, $J(k) = \frac{1}{2}e(k)e^*(k) = \frac{1}{2}[e_r^2 + e_i^2]$, which is a real function of a complex variables, is not analytic or differentiable in the complex sense, and does not satisfy the Cauchy–Riemann conditions [2].

The Cauchy–Riemann conditions therefore impose a very stringent structure on functions of complex variables, and several attempts have been made to introduce more convenient derivatives. This requires definition of a coordinate of complex functions in which the real and imaginary sectors will not need to be orthogonal.

6.7   Augmented Complex-Valued Matrix Variables

Since all the components of the two matrices $d\mathbf{Z}$ and $d\mathbf{Z}^*$ are linearly independent, we may define the augmented complex-valued matrix variable $\mathbb{Z}$ of size $N \times 2Q$, defined as follows:

$$\mathbb{Z} \triangleq [\mathbf{Z}, \mathbf{Z}^*] \in \mathbb{C}^{N \times 2Q}. \tag{6.9}$$

The differentials of all the components of $\mathbb{Z}$ are linearly independent [1]. Hence, the matrix $\mathbb{Z}$ can be treated as a matrix that contains only independent elements when finding complex-valued matrix derivatives. The main reason for introducing the augmented matrix variable is to make the presentation of the complex Hessian matrices more compact and easier to follow. When dealing with the complex matrix variables $\mathbf{Z}$ and $\mathbf{Z}^*$ explicitly, four Hessian matrices have to be found instead of one, which is the case when the augmented matrix variable $\mathbb{Z}$ is used. The differentials of the vectorization operators of the augmented matrix variables $\mathbb{Z}$ and its conjugate $\mathbb{Z}^*$ are given by

$$dvec(\mathbb{Z}) = \begin{bmatrix} dvec(\mathbf{Z}) \\ dvec(\mathbf{Z}^*) \end{bmatrix}.$$

$$dvec(\mathbb{Z}^*) = \begin{bmatrix} dvec(\mathbf{Z}^*) \\ dvec(\mathbf{Z}) \end{bmatrix}. \tag{6.10}$$

Using the above definitions many other complex matrix relations and derivations may be extracted [1]. As an example, in [66] separation of non-circular complex sources has been attempted. As an important conclusion of this article, compared to circular random variables, for non-circular random variables more statistics are available. This extra knowledge has been used in the context of BSS.

In the following section the concept of augmented statistics of complex-valued signals has been incorporated into the design of complex-valued SSA.

6.8   Singular Spectrum Analysis of P300 for Classification

As briefed in the early chapters of this book, event-related potentials (ERPs) are waveforms within EEGs as the brain responds to different brain audio, visual, or somatosensory stimulations. ERPs often have relatively smaller amplitudes compared with the background EEG, thus traditionally ERPs are elucidated using a signal-averaging procedure. Their source locations, strengths, latencies, or even their width provide important information for the clinical diagnosis of psychiatric diseases such as dementia, Alzheimer's or schizophrenia [56, 57].

Although EEGs offer fine-grained temporal resolution, they suffer from limited spatial resolution. Furthermore, some of the ERP components are likely to overlap, which makes it difficult to distinguish between specific components of the signal. A common example is the composite P300 wave, which is a positive ERP component with a latency of about 300 milliseconds after applying a task-relevant stimulus. P300 is distributed over the midline electrodes (Fz, Cz and Pz) and it generally has larger magnitude towards the parietal electrodes (positioned at the back of the head) [58]. The P300 consists of two major overlapping subcomponents known as P3a and P3b. P3a represents an automatic switch of attention to a novel stimulus regardless of the task, however, P3b is elicited by infrequent task relevant events. The P3b wave is mostly central-parietal while P3a has a frontal-central distribution and it is characterised by a shorter latency and more rapid habituation than P3b [56], [59]. Along with other clinical examinations, analysis of P300 could be used as a potential diagnostic procedure for a wide variety of brain disorders and a definite solution for various dementias, such as Alzheimer's and schizophrenia. For this purpose, a reliable method for separating P300 subcomponents must be employed. One of the most common methods for P300 detection is the traditional ERP averaging [57]. Averaging over a large number of trials could significantly enhance the P300 wave by reducing the background EEG, however, it suffers from several limitations. For example, by averaging it is assumed that P300 latencies are constant over the time while that is not the case in reality. In addition, this method ignores the effect of brain rapid habituation on P3a [60]. In other words, the person gets used to the stimuli. Thus, it is not easy to distinguish the small differences between subcomponents which are temporally overlapped [59]. In order to overcome these drawbacks and elucidate the P300 subcomponents, averaging can be applied over the smaller window frames with 50 temporal overlaps. Compared with the conventional averaging over all trials, this method does not reduce the background EEG significantly. Therefore, it would be beneficial to apply a robust smoothing algorithm to mitigate the effect of unwanted EEG while extracting the desirable P300 subcomponents.

Since SSA is a SVD-based algorithm, it has the potential to be used as a smoothing technique. This method can be extended to a novel augmented complex SSA (A-CSSA) to improve the extraction of P300 subcomponents. Complex-valued algorithms can take advantage of the correlation between two similar signals. It is not a surprise, therefore, that they have found many applications in machine learning and EEG analysis.

6.8.1 Augmented Complex SSA Algorithm

As stated before in this chapter, typically, the statistics of a complex domain are considered as the direct extension of real domain statistics. For example, the covariance matrix of a zero mean complex vector $\mathbf{f}$ can be formulated by replacing the standard transpose operator $(.)^T$ with the Hermitian transpose $(.)^H$, i.e. $\mathbf{ff}^T \rightarrow \mathbf{ff}^H$. However, recent works have shown that the basic complex covariance matrix ignores the correlation between the real and imaginary part of the signal; yet this information can be obtained using the pseudo-covariance matrix [2], [61].

To illustrate the correlation, consider a complex variable $z = a + jb$. Its covariance (c) can be calculated as $c = E[zz^*] = E[a^2 + b^2]$ and its pseudo-covariance (p) can be defined as $p = E[zz] = E[a^2 - b^2 + j2ab]$. The correlation is captured by the term $j2ab$.

Therefore, 'augmented' statistics have been established to generalise the optimal second order statistics for the complex domain [2] in which both covariance ($\mathbf{C}$) and pseudocovariance ($\mathbf{P}$) are considered:

$$\mathbf{f}_a = [\mathbf{f}, \mathbf{f}^*]^T \quad \rightarrow \quad \mathbf{C}_a = E[\mathbf{f}_a, \mathbf{f}_a^H] = E \begin{bmatrix} \mathbf{C} & \mathbf{P} \\ \mathbf{P}^* & \mathbf{C}^* \end{bmatrix} \qquad (6.11)$$

To incorporate the latest advances in complex-valued statistics into the CSSA framework, here we employed the augmented complex SSA (A-CSSA) [62].

The steps followed by the A-CSSA algorithm are summarised as follows [63];

Algorithm 1: Augmented complex SSA algorithm
During the decomposition stage we

1. Consider the input as a complex-valued vector $\mathbf{f}_s$ of length s,

2. Define the embedding dimension l as $1 < l < s$ and $n = s - l + 1$,

3. Generate a trajectory matrix $\mathbf{W}^{l,n}$ using the lagged version of the original signal,

4. Obtain the augmented version of the trajectory matrix $\mathbf{W}_a^{2l,n}$ by considering its conjugate,

5. Calculate the augmented covariance matrix $\mathbf{W}_a \mathbf{W}_a^H$,

6. Apply SVD to the generated covariance matrix and produce several eigentriple sets $(\lambda_j, \mathbf{q}_j, \mathbf{v}_j)$.

   During the reconstruction stage we

7. Select appropriate subgroups of the eigentriples based on the desirable output,

8. Generate the new trajectory matrix $\widetilde{\mathbf{W}_a}$ using only the selected eigentriples,

9. Reconstruct the desired complex-valued signal $\widetilde{\mathbf{f}}$ by hankelization.

As a remark, the new augmented covariance matrix in [10] for the A-CSSA, which includes both covariance and pseudo-covariance, takes into account the complete second order information in the complex-valued data. Thus, it is more likely that A-CSSA outperforms its basic complex-valued SSA for non-circular signals [62].

In addition, note that the A-CSSA method generates $2L$ eigenvalues which is double the size of its basic complex-valued counterpart. Hence, the higher computational complexity of ACSSA over CSSA is expected [62].

In an attempt by Enshaeifar et al. the A-CSSA algorithm has been applied to analyse the single trial ERP waveforms [63]. It has been reported that SSA smooths the data by grouping the dominant eigentriple sets. However, single-trial ERPs sometimes contain strong EEG background. In such a case, dominant eigen subspaces represent high power unwanted EEG, and smoothing cannot enhance the P300 wave. Therefore, in their proposed method they integrate a reference signal to emphasise the temporal location of P300 by empowering the corresponding eigentriples. To this end, each subject has its specific reference (say $\mathbf{x}$) calculated from the overall averaging of all the target events. This signal is then combined with a single-trial ERP (say $\mathbf{y}$) to construct a one-dimensional complex-valued vector $\mathbf{f}$ for the A-CSSA algorithm, that is, $\mathbf{f} = \mathbf{x} + j\mathbf{y}$.

Since augmented statistics takes into account the correlation between the real and imaginary parts of a signal, the reference signal from the real part of $\mathbf{f}$ enhances the P300 from dominant EEG background in the single-trial recordings. Furthermore, as such single-trials are measured for short time signal segments, it is very likely to detect and track the overlapping P300 subcomponents more accurately.

6.9   Experimental Results

The proposed complex-valued SSA was applied to 12 EEGs recorded for 8 healthy subjects and 4 schizophrenic patients during an auditory two-stimuli oddball experiment [62]. In the traditional oddball paradigm an infrequent target randomly occurs in a background of frequent standard stimuli and the subject is told to press a button when the target appears [58, 64].

All experiments were performed on the Cz channel. The main reasons for selecting Cz are: (i) Cz is central channel, so it contains the effect of both P3a and P3b which are distributed frontocentral and centroparietal respectively, (ii) as Cz is found on the central line, it can reflect the P300 wave even for abnormal cases with uni-lateral brain difficulties and (iii) the posterior alpha wave (8-13Hz) has slightly less effect on the central channel Cz.

The reference signal was obtained by temporally averaging over 35 target events which randomly appeared over a period of 320 seconds. On the other hand, single-trials were calculated over the moving window of 8 target stimuli with a 50% temporal overlap. All stimuli, including standards and targets, occurred every 2 seconds. However, in order to increase the chance of P3a generation for the first target, standard stimuli were repeated consecutively for about 50 seconds and the first target appeared three trials later to switch the subject's attention. The details of this process have been depicted in Figure 6.2.

The first segment (50–130sec) of the EEG signal from the C_Z channel for a schizophrenic patient is illustrated in Figure 6.3. The frame covers 80 seconds after the first target stimuli. As shown in the Figure, P3a and P3b are clearly visible using the proposed A-CSSA method (bottom-right). Following the same experiment, similar results were observed for 10 subjects.

In order to track the changes in P300 subcomponents, Figure 6.4 illustrates 250 seconds after the first target for a schizophrenic patient (top middle) and a healthy subject (bottom). Each row of the figure contains four subplots that represent different time frames. According to some prior knowledge, P300 is expected within first 500 milliseconds after the target onset. Hence, all the subplots are zoomed in for this range. Each subplot includes a complex-valued

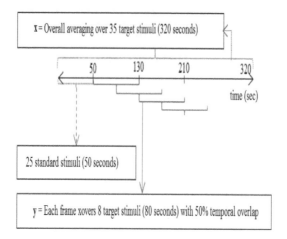

FIGURE 6.2: Each frame has an individual complex-valued signal generated as $f = x + jy$. Note that the overall average (y) is the same for all frames [63].

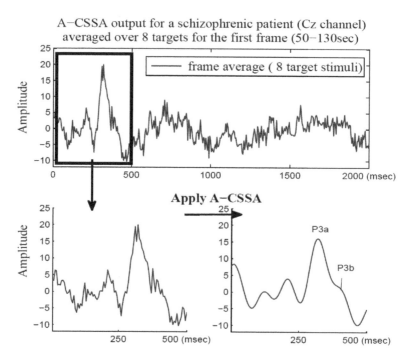

FIGURE 6.3: Original EEG segment for a schizophrenic patient recorded from central electrode Cz (top). The segment zoomed for 1–500msec after the target (bottom left). P300 subcomponents were clearly visible after using A-CSSA (bottom right) [63].

signal in which the real part (dashed line) is a reference signal obtained by averaging of all target stimuli and the imaginary part (solid line) is the average of 8 targets covering the period of 80 seconds. Note that each subject has a unique constant reference for all subplots.

6.10   Some Concluding Remarks

From the above study and the corresponding outputs the following conclusions may be derived [63]:

1. Figure 6.4 illustrates the rapid habituation of P3a over time after the first target event. This is the reason the overall average has less P3a amplitude than the frame averaged signal. see Figure 6.4 (b and c),

2. It is shown as the P3a habituates more, the P3b visibility improves. Generally, healthy subjects showed faster habituation than schizophrenic patients. Illustratively this is clear from Figure 6.4 (c) where P3b is stronger than P3a in the last temporal frame,

3. According to [65], significant reduction of P300 particularly in auditory experiments, is one of the most consistent biological findings in schizophrenia. This work was also in agreement with the literature and it is shown that P3b has constantly lower magnitude in schizophrenic patients (compare b and c),

4. In addition to the amplitude differences, schizophrenic patients showed longer P300 latency than healthy subjects.

The introduced A-CSSA is a new concept and algorithm. It is seen that A-CSSA may be employed to attenuate the EEG background by enhancing the ERP components. This certainly eases their tracking.

6.11   Extension to Hypercomplex Domain

The complex domain can be extended to a hypercomplex domain by means of quaternion signal processing and algebra.

Slightly different from a complex domain, a quaternion vector $\mathbf{x}$ has a real (scalar) part $\Re\{.\}$ and a vector part (also called pure quaternion)$\Im\{.\}$ which

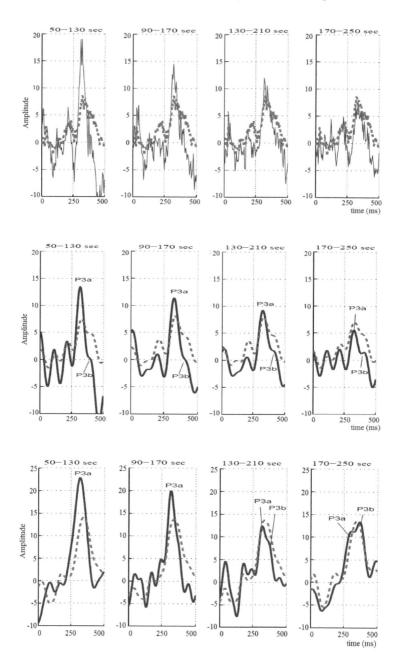

FIGURE 6.4: Application of A-CSSA for p3a-p3b detection; (a) original data for a schizophrenic patient, (b) highlighted P300 subcomponents (i.e. p3a and p3b) after A-CSSA for the same patient and (c) the components for a healthy subject. Temporal frame average is shown in bold and overall frame average in dotted lines [63].

consists of three imaginary components:

$$
\begin{aligned}
x &= \Re\{x\} + \Im\{x\} \\
&= \Re\{x\} + i\Im_i\{x\} + j\Im_j\{x\} + k\Im_k\{x\} \\
&= x_a + ix_b + jx_c + kx_d
\end{aligned}
\tag{6.12}
$$

Note that the imaginary units i, j and k are orthogonal unit vectors for which:

$$
ij = -ji = k
\tag{6.13}
$$
$$
jk = -kj = i
\tag{6.14}
$$
$$
ki = -ki = j
\tag{6.15}
$$
$$
i^2 = j^2 = k^2 = ijk = -1
\tag{6.16}
$$

The above identities demonstrate the non-commutative property of quaternion products where $x_1x_2 \neq x_2x_1$. To better extend the complex to hypercomplex algebra, another important notion for the quaternion domain (H), the so-called "quaternion involution" has to be studied. The involutions about the i, j and k imaginary unit axes are obtained from the components of quaternion as [67]:

$$
x^i = -ixi = x_a + ix_b - jx_c - kx_d
\tag{6.17}
$$
$$
x^j = -jxj = x_a - ix_b + jx_c - kx_d
\tag{6.18}
$$
$$
x^k = -kxk = x_a - ix_b - jx_c + kx_d
\tag{6.19}
$$

which form the bases for augmented quaternion statistics. Note that involution illustrates a rotation along a single unit axis, while the quaternion conjugate operator $(.)^*$ rotates along all three imaginary axes:

$$
x^* = \Re x - \Im x = x_a - ix_b - jx_c - kx_d
\tag{6.20}
$$

The relationship between a quaternion vector and its involutions is given by:

$$
x^* = \frac{1}{2}[x^i + x^j + x^k - x]
\tag{6.21}
$$

From this equation as soon as the quaternion involutions are known the conjugate of a quaternion can be computed.

In order to achieve an augmented covariance matrix such as the one for complex-valued statistics, the correspondence between the elements of a quaternion variable in **H** (Hamiltonian domain) and the elements of a quadri-variate vector in $\mathbb{R}^4$ can be obtained in [67]:

$$
x_a = \frac{1}{2}[x + x^*], \quad x_b = \frac{1}{2}[x - x^{i*}]
\tag{6.22}
$$

$$x_c = \frac{1}{2}[x - x^{j*}], \quad x_d = \frac{1}{2}[x - x^{k*}] \qquad (6.23)$$

Thus, the quaternion statistics should include all quaternion involutions x_i, x_j and x_k to access to the complete second order statistical information. Following the above derivations, the augmented quaternion vector $x^a = [x^T; x^{iT}; x^{jT}; x^{kT}]^T$ is defined and used to compute the augmented covariance matrix derived as:

$$C_q^a = E[x^a x^{aH}] = \begin{bmatrix} C_{xx} & C_{x^i} & C_{x^j} & C_{x^k} \\ C_{x^i}^H & C_{x^i x^i} & C_{x^i x^j} & C_{x^i x^k} \\ C_{x^j}^H & C_{x^j x^i} & C_{x^j x^j} & C_{x^j x^k} \\ C_{x^k}^H & C_{x^k x^i} & C_{x^k x^j} & C_{x^k x^k} \end{bmatrix}$$

In the above augmented covariance matrix, the term $C_{xx} = E[xx^H]$ is the standard covariance matrix, whereas other terms are called complementary or pseudo-covariance matrices and are defined as:

$$C_{x^\alpha} = E[xx^{\alpha H}], \text{ for all } \alpha\beta = i, j, k \qquad (6.24)$$

$$C_{x^\alpha x^\beta} = E[x^\alpha x^{\beta H}] \qquad (6.25)$$

In an application by Enshaeifar et al. [68] the above augmented covariance matrix has been used to estimate a quaternion SVD of an augmented quaternion SSA (AQSSA). In this application four EEG channel signals have been considered as the real and imaginary components of a quaternion-valued signal for study of sleep EEG. The EEG electrodes used in this scenario are circled in Figure 6.5.

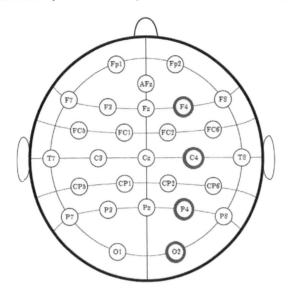

FIGURE 6.5: positions of four electrodes used for sleep study using quaternion domain SSA.

Being more specific, in this study, it has been shown that the AQSSA may be used for three or four dimensional/channel signal processing. As an elegant work, the signals from four EEG channels have been used for sleep scoring. The significance of such approach is that the spatial diversity is somehow exploited to ensure that all the signal details with sources in different brain zones are captured and effectively combined [68].

The approach in AQSSA determines the grouping of the selected elementary matrices by measuring the correlation between the reconstructed signal of the desired frequency bands. That allowed for all the sleep stages to be detected in an automated way.

In the context of sleep EEG, the above AQSSA method is used to determine the statistical descriptors for a 5-state (Awake, N1, N2, N3 and REM) sleep classification.

Stage N1 is the drowsiness stage that is considered as a transition stage from awake to asleep. In this stage, the alpha rhythm is attenuated and replaced by low amplitude waves, predominantly 4-7 Hz or theta, for more than 50% of the epoches. The next stage or Stage N2 is characterised by the occurrence of sleep spindles or K complexes. Stage N2 might lead to Stage N3 which consists of slow wave activities (0.5–2 Hz or delta) measured frontally in adults. The spindles may exist in this stage, however, as the sleep deepens the spindles' rate decreases. Using these characteristics, the quaternion SSA can be used to estimate five brain rhythms to track the NREM and REM stages of sleep.

Application of AQSSA provided 74% epoch agreement between the automated and manual scoring obtained in the Sleep Centre of University of Surrey, UK. Certainly, the method has the potential to be used in various multivariate biomedical applications, such as BCI or gait analysis for rehabilitation purposes [68].

6.12   Conclusion

Complex and hypercomplex signal processing involving the new concepts of augmented statistics and the development of widely linear models, which then need to be optimised, have opened a new front in multi-dimensional and multichannel signal processing. Naturally, as the consequence, SSA can be applied to two or four channel signals respectively while unlike many source separation techniques the signals are considered to be correlated or dependent.

Bibliography

[1] Hjorungnes, A. (2011). *Complex-Valued Matrix Derivatives*. Cambridge University Press.

[2] Mandic, D. P., and Goh, V. S. L. (2009). *Complex valued nonlinear adaptive filters: noncircularity, widely linear and neural models*. John Wiley & Sons.

[3] Schreier, P. and Scharf, L. (2010). *Statistical Signal Processing of Complex-Valued Data: the Theory of Improper and Noncircular Signals*. Cambridge University Press, Cambridge, UK.

[4] Paulraj, A., Nabar, R., and Gore, D. (2003). *Introduction to Space-Time Wireless Communications*. Cambridge University Press, Cambridge, UK.

[5] González-Vázquez, F. J. (1988). The differentiation of functions of conjugate complex variables: Application to power network analysis. *Education, IEEE Transactions*, **31**(4), 286–291.

[6] Alexander, S. T. (1984). A derivation of the complex fast Kalman algorithm.Acoustics. *Speech and Signal Processing, IEEE Transactions*, **32**(6), 1230–1232.

[7] Hanna, A. I., and Mandic, D. P. (2003). A fully adaptive normalized nonlinear gradient descent algorithm for complex-valued nonlinear adaptive filters. *Signal Processing, IEEE Transactions*, **51**(10), 2540–2549.

[8] Han, Z. and Liu, K. J. R. (2008). *Resource Allocation for Wireless Networks: Basics, Techniques, and Applications*. Cambridge University Press, Cambridge, UK.

[9] Franken, D. (1997). Complex digital networks: A sensitivity analysis based on the Wirtinger calculus. *Circuits and Systems I: Fundamental Theory and Applications, IEEE Transactions*, **44**(9), 839–843.

[10] Kreyszig, E. (1988). *Advanced Engineering Mathematics*, 6th ed., John Wiley and Sons, New York, USA.

[11] Boyd, S. and Vandenberghe, L. (2004). *Convex Optimization*. Cambridge University Press, Cambridge, UK.

[12] Palomar, D. P. and Eldar, Y. C. (Eds.). (2010). *Convex Optimization in Signal Processing and Communications.* Cambridge University Press, Cambridge, UK.

[13] Dwyer, P. S. and Macphail, M. S. (1948). Symbolic matrix derivatives. *Annals of Mathematical Statistics,* **19**(4), 517–534.

[14] Dwyer, P. S. (1967). Some applications of matrix derivatives in multivariate analysis. *Journal of the American Statistical Association,* **62**(318), 607–625.

[15] Nel, D. G. (1980). On matrix differentiation in statistics. *South African Statistical Journal,* **14**(2), 137–193.

[16] Wirtinger, W. (1927). Zur formalen Theorie der Funktionen von mehr komplexen Veranderlichen. *Mathematische Annalen,* **97**(1), 357–375.

[17] Brandwood, D. H. (1983). A complex gradient operator and its application in adaptive array theory. In *IEE Proceedings F (Communications, Radar and Signal Processing),* **130** (1), 11–16. IET Digital Library.

[18] Graham, A. (1981). *Kronecker Products and Matrix Calculus with Applications.* Ellis Horwood Limited, England.

[19] Moon, T. K. and Stirling, W. C. (2000). *Mathematical Methods and Algorithms for Signal Processing.* Prentice Hall, Englewood Cliffs, NJ, USA.

[20] Trees, H. L. V. (2002). *Optimum Array Processing: Part IV of Detection Estimation and Modulation Theory.* Wiley Interscience, New York, USA.

[21] Magnus, J. R. and Neudecker, H. (1988). *Matrix Differential Calculus with Application in Statistics and Econometrics.* John Wiley and Sons, Essex, UK.

[22] Harville, D. A. (1997). *Matrix Algebra from a Statistician's Perspective.* Springer-Verlag, New York.

[23] Minka, T. P. (2000). *Old and new matrix algebra useful for statistics.* [Online Available]: www. stat. cmu. edu/minka/papers/matrix. html.

[24] Lutkepohl, H. (1996). *Handbook of Matrices.* John Wiley and Sons, New York, USA.

[25] Hjørungnes, A., and Gesbert, D. (2007). Complex-valued matrix differentiation: Techniques and key results. *Signal Processing, IEEE Transactions,* **55**(6), 2740–2746.

[26] Van Den Bos, A. (1994). Complex gradient and Hessian. *IEE Proceedings-Vision, Image and Signal Processing,* **141**(6), 380–382.

[27] Hjørungnes, A. and Gesbert, D. (2007b). Hessians of scalar functions which depend on complex valued matrices. In *Proc. Int. Symp. on Signal Proc. and Its Applications*, Sharjah, United Arab Emirates.

[28] Hjørungnes, A. and Palomar, D. P. (2008). Patterned complex-valued matrix derivatives. In *Proc. IEEE Int. Workshop on Sensor Array and Multi-Channel Signal Processing*, Darmstadt, Germany, 293–297.

[29] Hjørungnes, A. and Palomar, D. P. (2008). Finding patterned complex-valued matrix derivatives by using manifolds. In *Proc. Int. Symp. on Applied Sciences in Biomedical and Communication Technologies*, Aalborg, Denmark.

[30] Palomar, D. P., and Verdú, S. (2006). Gradient of mutual information in linear vector Gaussian channels. *IEEE Transactions of Information Theory*, **52**(1), 141–154.

[31] Vaidyanathan, P. P., Phoong, S. M., and Lin, Y. P. (2010). *Signal Processing and Optimization for Transceiver Systems*. Cambridge University Press, Cambridge, UK.

[32] Priestley, I. A. (2003). *Introduction to Complex Analysis*. Oxford University Press.

[33] Widrow, B., McCool, J., and Ball, M. (1975). The complex LMS algorithm. In *IEEE Proceeding*, **63**, 719.

[34] Picinbono, B., and Chevalier, P. (1995). Widely linear estimation with complex data. *IEEE Transactions on Signal Processing*, **43**(8), 2030–2033.

[35] Brown, W. M., and Crane, R. B. (1969). Conjugate linear filtering. *Information Theory, IEEE Transactions*, **15**(4), 462–465.

[36] Van Den Bos, A. (1995). The multivariate complex normal distribution: A generalization. *IEEE Transactions on Information Theory*, **41**(2), 537–539.

[37] Neeser, F. D., and Massey, J. L. (1993). Proper Complex Random Processes with Application to Information Theory. *IEEE Transactions on Information Theory*, **39**(4), 1293–1302.

[38] Schreier, P. J., and Scharf, L. L. (2001, November). Low-rank approximation of improper complex random vectors. In *Signals, Systems and Computers, 2001. Conference Record of the Thirty-Fifth Asilomar Conference*, **1**, 597–601.

[39] Schreier, P. J., and Scharf, L. L. (2003). Second-order analysis of improper complex random vectors and processes. *Signal Processing, IEEE Transactions*, **51**(3), 714–725.

[40] Schreier, P. J., Scharf, L. L., and Mullis, C. T. (2005). Detection and estimation of improper complex random signals. *Information Theory, IEEE Transactions*, **51**(1), 306–312.

[41] Picinbono, B. (1994). On circularity. *Signal Processing, IEEE Transactions*, **42**(12), 3473–3482.

[42] Picinbono, B., and Chevalier, P. (1995). Widely linear estimation with complex data. *IEEE transactions on Signal Processing*, **43**(8), 2030–2033.

[43] Goh, S. L., and Mandic, D. P. (2007). An augmented extended Kalman filter algorithm for complex-valued recurrent neural networks. *Neural Computation*, **19**(4), 1039–1055.

[44] Goh, S. L., and Mandic, D. P. (2007). An augmented CRTRL for complex-valued recurrent neural networks. *Neural Networks*, **20**(10), 1061–1066.

[45] Calhoun, V., Adali, T., and Van De Ville, D. (2009). *The Signal Processing Handbook for Functional Brain Image Analysis*. Springer.

[46] Calhoun, V. D., Adali, T., Pearlson, G. D., Van Zijl, P. C. M., and Pekar, J. J. (2002). Independent component analysis of fMRI data in the complex domain. *Magnetic Resonance in Medicine*, **48**(1), 180–192.

[47] Eriksson, J., and Koivunen, V. (2006). Complex random vectors and ICA models: Identifiability, uniqueness, and separability. *Information Theory, IEEE Transactions*, **52**(3), 1017–1029.

[48] Roman, T., Visuri, S., and Koivunen, V. (2006). Blind frequency synchronization in OFDM via diagonality criterion. *Signal Processing, IEEE Transactions*, **54**(8), 3125–3135.

[49] Gautama, T., Mandic, D. P., and Van Hulle, M. M. (2004). A nonparametric test for detecting the complex-valued nature of time series. *International Journal of Knowledge-based and Intelligent Engineering Systems*, **8**(2), 99–106.

[50] Schreier, P. J., Scharf, L. L., and Hanssen, A. (2006). A generalized likelihood ratio test for impropriety of complex signals. *Signal Processing Letters*, **13**(7), 433–436.

[51] Schreier, P. J., and Scharf, L. L. (2010). *Statistical Signal Processing of Complex-Valued Data: the theory of improper and noncircular signals*. Cambridge University Press.

[52] Needham, T. (1997). *Visual Complex Analysis*. Oxford University Press.

[53] Vaucher, G. (2003). A complex valued spiking machine. In *Proceedings of ICANN 2003*, 967–976.

[54] Vaucher, G. (2003). A complex valued spiking machine. In *Proceedings of ICANN 2003*, 967–976.

[55] Aizenberg, I., Myasnikova, E., Samsonova, M., and Reinitz, J. (2002). Temporal classification of Drosophila segmentation gene expression patterns by the multi-valued neural recognition method. *Mathematical Biosciences*, **176**(1), 145–159.

[56] Sanei, S. (2013). *Adaptive Processing of Brain Signals*. John Wiley and Sons.

[57] Spyrou, L., Jing, M., Sanei, S., and Sumich, A. (2007). Separation and localisation of P300 sources and their subcomponents using constrained blind source separation. *EURASIP Journal on Applied Signal Processing*, bf 1, 89–89.

[58] Polich, J. (2007). Updating P300: an integrative theory of P3a and P3b. *Clinical neurophysiology*, **118**(10), 2128–2148.

[59] Spyrou, L., and Sanei, S. (2006). A Robust Constrained Method for the Extraction of P300 Subcomponents. In *Acoustics, Speech and Signal Processing. ICASSP 2006 Proceedings. 2006 IEEE International Conference*, **2**, (II).

[60] Ford, J. M. (1999). Schizophrenia: the broken P300 and beyond. *Psychophysiology*, **36**(6), 667–682.

[61] Schreier, P. J., and Scharf, L. L. (2010). *Statistical signal processing of complex-valued data: the theory of improper and noncircular signals.* Cambridge University Press.

[62] Enshaeifar, S., Sanei, S., and Took, C. C. (2014). An eigen-based approach for complex-valued Forecasting. In *Acoustics, Speech and Signal Processing (ICASSP), 2014 IEEE International Conference*, 6014–6018.

[63] Enshaeifar, S., Sanei, S. and Cheong Took, C. (2014). Singular Spectrum Analysis of P300 for Classification. *Proc. of WCCI*, China.

[64] Polich, J., and Criado, J. R. (2006). Neuropsychology and neuropharmacology of P3a and P3b. *International Journal of Psychophysiology*, **60**(2), 172–185.

[65] Park, E. J., Han, S. I., and Jeon, Y. W. (2010). Auditory and visual P300 reflecting cognitive improvement in patients with schizophrenia with quetiapine: a pilot study. *Progress in Neuro-Psychopharmacology and Biological Psychiatry*, **34**(4), 674–680.

[66] De Lathauwer, L., and De Moor, B. (2002). On the blind separation of non-circular sources. Mh, **100**(1), 1.

[67] Took, C. C., and Mandic, D. P.(2011). Augmented second-order statistics of quaternion random signals. *Signal Process*, **91**(2), pp. 214–224.

[68] Enshaeifar, S., Kouchaki, S., Took, C.C., and Sanei, S. (2014). Quaternion singular spectrum analysis for underdetermined source separation with application to sleep data. *IEEE Trans. on Neural Systems and Rehabilitation Engineering*, Forthcoming.

Chapter 7

SSA Change Point Detection and Eye Fundus Image Analysis

In parallel with development of high speed complex computing systems and software tools, automated image analysis has become popular in various fields such as security, biometrics, medicine, navigation, and art. In addition, machine learning has been actively researched to fulfil the above demands and enable interpretation of the image contents.

For assessing digital retinopathy images such image processing tools become very advantageous. For better scalability, the researchers have been developing new techniques to improve the performance of their algorithms.

Image enhancement, image restoration, segmentation of image regions including edge detection, and even texture analysis techniques have been under intensive research. The objective of this chapter however, is not to provide any overview of image processing techniques but to describe how the change point detection capability of SSA allows enhancement or identification of blood vessels and capillaries within eye fundus images.

There have been many approaches for detection of vessels from such images mainly through image segmentation, edge detection, and curve tracing recently [7]. Unfortunately, they all have had limited success and poor performance improvement [7].

As the fundamental objective of these algorithms the precise localisation of the normal or abnormal anatomic structure, and the difficulty to understand the variability of the image content as well as the similarity of clinical signs to other part of retinal images can be mentioned. As an example, many existing algorithms fail to accurately detect the red lesions such as microaneurysms and small haemorrhages.

As part of the overall fundus image processing algorithm, a suitable means of image quality assessment has to be established. The images should also be normalised to alleviate the effect of shading and change of light. Then, the non-uniform illumination should be corrected. The colours also should be normalised since colour is a significant descriptor and a feature used to distinguish between different retinal abnormalities. The fluorescein-labelled images, 'red-free' images, and the green channel of RGB are widely used as the features for automatically detecting the retinal abnormalities. Very often other pre-processing techniques, such as contrast enhancement, are also applied.

We see later on in this chapter, as an important application of SSA, that change point detection is indeed useful in detection of boundaries and segmentation of various data modalities. A sudden change in the signal trend may be easily detected by subtraction of a smoothed version of the data (by employing SSA) from the data itself. This is equivalent to the gradient of the data. Using SSA with properly selected parameters however, the effect of noise and other artefacts can also be mitigated or removed. In this chapter the objective is to exploit this property in order to detect the blood vessels within eye fundus images and use the achieved information for diagnostic purposes.

7.1   Ocular Fundus Abnormalities

Much research has been undertaken for detection and monitoring of ocular diseases within the past two decades. This is partly due to the increase in the number of people of all ages living with sight loss. This number is predicted to reach approximately 285 million, with 39 million blind [2].

Currently, around 90% of visually impaired people in the world are in developing countries, and 1.4 million children around the world are irrevocably blind for the rest of their lives. In the United Kingdom, there are almost 2 million people living with sight loss that has a momentous impact on their daily life [3]. However, 80% of all vision loss can be prevented or cured. The prevention, management and treatment of the ocular diseases is important, people with vision loss will feel 'moderately' or 'completely' cut of from people and things around them.

It is evident that population groups, either defined by geographical areas or ethnicity, may exhibit distinctive patterns with regard to the symptoms and manifestation of ocular illnesses and referrals to the related clinical departments [4].

The ocular fundus is the inner lining of the eye made up of the sensory retina, the retinal pigment epithelium, bruch's membrane, and the choroid.

Prevalence of moderate or serious retinopathy is 4% among Europeans, 12.9% among the Maori and 15.8% among the Pacific populations. Cataracts can be found in 19.3% of Europeans, 36.6% of the Pacific population and 16.4% of the Maori ethnic group [4].

Especially in the western world, diabetic retinopathy (DR) is the leading cause of blindness in the working-age population. While DR can happen to anyone with diabetes, certain ethnic groups are at higher risk because they are more likely to have diabetes, for example, African Americans, Latinos and native Americans. Some ethnic groups (including tribal populations) are two to four times more susceptible to diabetes; as a result, the rates of DR may represent the rates of diabetes. In addition, poor access to health care remains a significant problem in some ethnic groups and in those who are socially and economically deprived. Screening and prompt treatment of DR are not top priorities in many countries of the world, due to the fact that the impacts of other causes of preventable blindness remain an issue [6]. Hence, it is interesting to have a population-based study to understand the differences and possible patterns in ocular disease.

7.2   Ocular Fundus Images

Like other diseases, one of the efficient ways to prevent and manage the ocular diseases is the regular check-ups or diagnosis of any eye problems. With the development of ophthalmic photography, an ophthalmologist can diagnose a patient based on the medical images obtained from ophthalmic photography. The fundus camera has a powerful lens that can penetrate through the pupil (sometimes the pupil has to be dilated), through the Vitreous gel (the gel that is inside the eye) and photograph the retina (the back of the eye) [6]. The doctor can set up the fundus camera according to the kind of the diagnosis. For instance, for patients that have or might have diseases around the optic nerve, for example glaucoma, a stronger magnified lens is used to focus on the optic nerve. In the UK national diabetic retinopathy screen programme, each patient will have two images taken for each eye, that is, there are four images for each person. Given there are nearly 3 million diabetes patients in the UK, a huge number of digital images is created for check-ups each year.

The fundus can be examined using an ophthalmoscope and/or fundus photography. Figure 7.1 demonstrates a fundus of human eye.

Furthermore, with the development of ophthalmic photography devices, the medical photos are easily captured using digital cameras at people's homes and work, and the image data can be transferred to be read and identified by the ophthalmologist.

Research has been carried out to develop a device which allows people to capture suitable pictures of the optic fundus and retinal nerve using a smart phone [10]. It is obvious that the workload for the ophthalmologist will be substantial, if all ocular disease patients are offered regular screening with ophthalmic photography.

Another fact is that in DR screening, most of the screening images, about 2/3, are normal. Therefore, it order to reduce the workload of the ophthalmol-

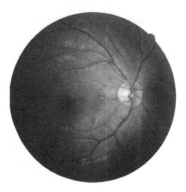

FIGURE 7.1: A fundus photo, showing the optic disc as a bright area on the right where blood vessels converge. The spot to the left of the centre is the macula. The grey, more diffuse spot in the centre is a shadow artifact.

ogist, it is necessary to develop an automatic system to process the normal medical images with high accuracy before its inspection for diagnostic purposes by the ophthalmologists.

The human visual system has no problem representing information such as shape, shading and translucency from the different environments. Over the past decades, the perceptual psychologists have tried to understand how human vision works; meanwhile, researchers have attempted to provide the perceptual capability for a machine which is able to duplicate humans' understanding of images. Computer vision is an area of research that aims for a machine to see and understand an image by imitating the effect of human visual system. It has been infiltrating deeper and wider into the various domains such as medical image processing, machine vision, and military and autonomous vehicles. However, providing computers with the ability to perceive the effective information is a complicated task. It needs to combine various disciplines including image processing, communications, psychology, mathematics, pattern recognition, artificial intelligence, etc.

DR is the damage to the retina caused by complications of diabetes, which can eventually lead to sight loss [3]. Diagnosis of retinal images for the presence of DR can be seen as a pattern recognition process. The typical clinical signs of DR are red lesions: microaneurysms, small dot haemorrhages, larger blotchy haemorrhages; and bright (or white) lesions: lipid exudates, cotton wool spots, etc. Ophthalmologists or trained graders need to recognise these red and bright lesions from the background of retinal images. However, recognising clinical features and evaluating their relationship in retinal images require a series of processing for machines.

These may involve pre-processing, image segmentation, feature extraction and classification. Pre-processing can be applied for correcting non-uniform illumination, colour normalisation, noise reduction or contrast enhancement.

Using segmentation we try to detect the boundaries in order to separate the objects from the image background or from each other, including retinal blood vessels, red and white lesions or optic disc.

Feature extraction is to obtain information from the segmented objects such as their size, colour or morphological form. According to some previously established criteria, the segmented objects will be classified using the extracted features. This is, however, a typical pattern recognition process. These types of images are often difficult to analyse since:

1. The diagnostic signs and symptoms within abnormal eye images have irregular appearance, random placement and unpredictable quantity.

2. Different abnormalities may manifest visually similar features. For example, large clusters of exudates can be regarded as an optic disc or clinical signs caused by other ocular disease due to their similar brightness.

3. Illumination, acquisition angle, or retinal pigmentation can affect the attributes of retinal features.

4. The subtle clinical changes are usually primary evidence of disease progression, however, their subtlety is extremely challenging for computer vision.

Due to these issues, it is a great challenge to accurately assess the clinical signs in retinal images. Meanwhile, this task needs to achieve good *sensitivity* and *specificity* in order to be useful in clinical practice.

7.3 Diabetic Retinopathy Images

As one of the most prevalent eye diseases around the globe, diabetic retinopathy, is the main cause of vision loss among working-age population. An early regular checkup of the patients with diabetes is advised to best prevent the loss of vision. The UK National Screening Committee recommends that all people with diabetes aged 12 and over should be screened every year for DR using digital retinal photography [3]. The prevalence of diabetes mainly originates from population growth, an ageing population, unhealthy diets, physical inactivity and a rising incidence of obesity [2]. The global population of diabetes is estimated to rise from 366 million in 2011 to 552 million by 2030 [2]. By 2025 it is likely that approximately 5 million people will suffer from diabetes in the UK [3]. On the other hand, screening the entire population implies a high clinical workload.

Retina images for diabetic retinopathy screening are acquired through a non-intrusive method using digital fundus cameras [3]. Figure 7.2 illustrates some typical retinopathy images which have the same anatomic structure but varying brightness, contrast, and appearance.

There are many factors which result in the variability and diversity of the images, such as dust, eyelashes, patient's ethnic origins and eye blink. These situations make the processing of such images a difficult task [9].

A healthy retina image contains three main components: the optic disc, blood vessels and the macula (Figure 7.3). The right and left eye can be recognised by the relative location of the optic disc and the macula. Detecting these anatomical structures requires a low probability of false positives. For example, the end of the blood vessels is often mistaken for microaneurysms due to their similar visual features measured during image processing.

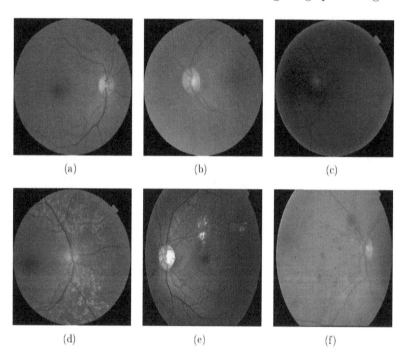

(a) (b) (c)

(d) (e) (f)

FIGURE 7.2: various typical retinopathy images.

Optic Disc: The optic disc (OD) is an oval spot on the back of the eye made up of nerve cells. Generally, the optic disc is brighter than the surrounding area in the retinal fundus images. However, variations of pigmentation in health eyes can cause differences in the appearance of the disc. Main branches of the central retinal blood vessel artery run across the disc contour. Figure 7.4 (a) shows the bright lesion that might cause confusion in detection of the OD, and also shows that the OD is not the brightest region.

Macula: The macula is a spot in the retina that surrounds the fovea. The

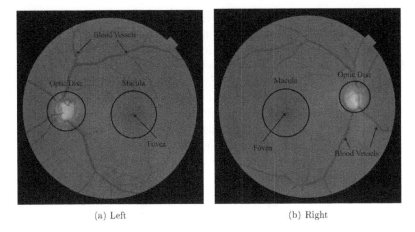

(a) Left   (b) Right

FIGURE 7.3: Anatomic structures of retina for left and right eyes.

clinical signs appear around this location, which is regarded as a threatening condition. The macula is localised about 2–2.5 disc diameters temporal to the edge of the OD and between the major temporal retinal vessels [9]. When some clinical signs like microaneurysms appear in this region, it is often very hard to detect the signs as both being dark, the contrast between them is low. This can be seen in Figure 7.4 (b).

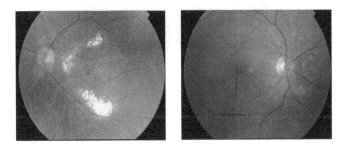

FIGURE 7.4: Bright lesion (a) and microaneurysms (b) [9].

Blood Vessels: The retinal vasculature includes the arteries and veins, with their branches. The central retinal artery bifurcates at or on the OD into divisions that supply the four quadrants of the inner retinal layers. A similarly arranged system of retinal veins intersects at the OD [10].

Relative to other retinal surfaces, the blood vessels have a lower reflectance. As a result, the blood vessels appear darker compared to the background. In an unhealthy retinal image, the tortuosity of blood vessels can change dramatically. Since there is very low contrast between those subtle tiny vessels and the retinal background, they are very challenging to detect without comprehensive analysis techniques. Figure 7.5 (a) shows an example of the blood vessel being distorted and not clearly visible due to a severe retinopathy. Figure 7.5 (b) illustrates an example of subtle vessels and their fragile response to a simple image segmentation technique.

Detection of the retinal blood vessels is an important stage of the process for diagnosis and treatment monitoring of fundus abnormality. Therefore, accurate tracing of the retinal blood vessels contributes to pathological diagnosis of the diseases. Moreover, the segmentation of blood vessels plays an important role in registration and spatial alignment of fundus images.

The blood vessels may appear darker (as they often do) or lighter than the background eye pattern in some other cases [10].

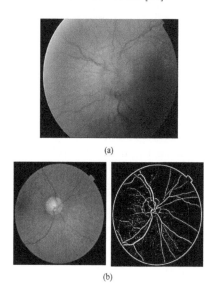

(a)

(b)

FIGURE 7.5: (a) Distorted blood vessels and (b) blood vessel segmentation for a reasonable quality image.

In the retinal images, different clinical signs can occur at any point. An ophthalmologist or a qualified screening technician would assess retinopathy severity by detecting the different DR signs. The typical clinical signs of DR are red (or dark) lesions (microaneurysms, small dot haemorrhages, larger blot haemorrhages) and white (bright) lesions, for example lipid exudates, and cotton wool spots. Trained graders need to recognise these red and white lesions from the background of retinal images.

7.4 Analysis of Fundus Images and Detection of Retinopathy Features

Many automatic approaches for vessel segmentation have been proposed [12]–[19], however, there is an understanding of their scalability across images from diverse populations. Typical images obtained from one population usually present different quality, contrast and retinal pigment background. When studying images acquired from various populations, new patterns such as Retinitis Pigmentosa may also appear, which may add further complexity in understanding the vessel structures using the existing methods [20]. Therefore, visual inspection of such images is a good way of diagnosing the diseases and abnormalities. Our motivation is to seek a reliable approach to generate vessel networks on very large scale data across diverse populations.

Recently, there has been a popular trend to utilise supervised approaches to extract the vessels including k-nearest network, support vector machines, feature-based Gradient Boosting methods, etc. [12]–[14]. These require samples of the ground truth for training the models, which is not easily available. The ground truth of retinal vascular structures based on a small number of images was manually annotated by ophthalmologists. However, the 're-training' process is always required on new datasets when further manual annotations are not easily acquired. Nevertheless, supervised approaches can generate higher accuracy than other methods for healthy retinal images where no dark lesions are present.

Another proposed approach is based on the matched filtering methodology to enhance the blood vessels in fundus images [15]. This approach assumes that the intensity profile of a cross-sectional retinal vessel approximates a Gaussian curve and the vessel diameter gradually decreases as it moves away from the optic disc. The matched filter was first proposed in [15], and different variants have been investigated to improve its performance. For example, the original matched filter is adapted and extended with the first order derivative of Gaussian image to detect many fine vessels and reduce false detection of non-vessel structures [16].

Due to these naive assumptions, the main disadvantage of matched filter methods is that they are very difficult to adapt to the variations in vessel width and orientations. In addition, the presence of a central reflex and the variations of retinal background in the retinal images may cause a number of false responses.

The approaches based on the vessel tracking methodology segment a vessel between two points by measuring local information. Chutatape et al. proposed a tracking strategy using a Gaussian and Kalman filter to trace blood vessels in retinal images [17]. Can et al. exploited directional templates to recursively track the vessels starting from initial seeds [18]. Some semi-automated methods for segmentation of vessel networks are proposed by Wang et al. [19]. The main advantage of tracking based methods is that the morphological features are accurately extracted during tracking. In addition, tracking methods are efficient since they do not search some area of the retinal image which does not include vessels.

The problem of such approaches is that vessel tracking algorithms required initial seeds to trace the vessel networks and may miss some of the bifurcation or crossover points, and thus cannot extract sub-trees.

On the other hand, in some applications, exudates are treated separately. In the retinal images, exudates appear as yellow objects of irregular shapes and sizes. The challenges of exudates detection lie in the fact that they are not the only bright objects in the retinal image. The bright lesions have very similar shapes, sizes, and colours. This may be seen when comparing the optic disc, cotton wool spot, and drusen (drusen results from accumulation of hyaline particles underneath the retina).

Sinthanayothin et al. proposed a similar recursive region-growing method to extract the exudates in colour fundus images [12]. As a consequence, the exudates were overlaid onto the original fundus image. According to 30 retinal images (21 images with exudates), they declared an 88.5% sensitivity and 99.7% specificity. Walter et al. used the blood vessels' removal based on the morphological operations and then exudates' regions extracted by candidate regions were subtracted from the image which remove blood vessels [1]. They reported 92.8% sensitivity and 92.4% specificity for 30 images. Gardner et al. reported a 93.1% sensitivity using methods that relied on neural networks [11]. Philip et al. used a multi-scale morphological operation to extract the candidate regions for exudates [16]. Then local properties were applied to classify the exudates, drusens and background. In 13219 images, they achieved respectively 95% and 84.6% of sensitivity and specificity. Although these methods for detecting exudates achieve high accuracies, their dataset is not inclusive and doesnt include all scenarios. Their study only involved segmentation of bright lesions from dark lesions without distinguishing the exudates, optic disc, and cotton wool spots.

In a recent work by Wang et al. [9], a method to construct the vessel networks has been proposed which addresses the above problems. In this work first a set of partitioned vessel segments is extracted by using SSA to generate a vessel distribution map. As the initial seeds, the vessel distribution map is used to track the blood vessels. Then, the map is traced in the reverse direction to recover the lost segments belonging to the capillaries.

7.4.1 Potential Vessel Distribution Map

The potential vessel distribution map (PVDM) consists of a set of potential vessel fragments. On a PVDM, each pixel is an initial seed which is applied to trace the neighbouring vessels and construct the vessel networks. Figure 7.6 is an example of the resulting PVDM. For obtaining a PVDM, we adopt the basic SSA procedure to decompose a retinal image into a number of components including slowly varying trends, oscillatory components and unstructured noise. The algorithm has been suggested in [25]. An original retinal image is transformed to two 1-D directional series through both horizontal and vertical directions (Figure 7.6 (b) and (c)) in order to retain all the information about all the neighbourhoods of the pixel at (s_x, s_y) of length r. Then, each series is converted to a trajectory matrix $\mathbf{X}$ following the routine in SSA. The trajectory matrix $\mathbf{X}$ is a Hankel matrix which has equal elements for all the diagonals $i + j = constant$.

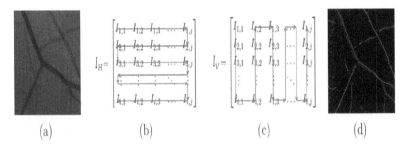

(a)   (b)   (c)   (d)

FIGURE 7.6: Construction of PVDM by employing SSA; (a) portion of fundus image in gray channel, (b) and (c) transformation from 2-D image into 1-D series in horizontal s_x and vertical direction s_y, and (d) the final PVDM in horizontal and vertical directions.

In the next stage, a conventional SSA algorithm is followed to decompose the covariance of the Hankel matrix into the set of eigentriples.

In SSA application, a priori knowledge is generally used to reject the noise subspace and extract the trends or particular periodicities. By application of SSA we can decompose the original data into a number of principal components. These components fall into either signal or noise subspace. Here, the objective is not to separate different components from the time series. Instead, the aim is to identify the subspace that the desired signal components belong to. As a consequence, the largest eigenvalues belong to the signal subspace and the small eigenvalues belong to that of noise.

After removing the noise subspace, the new matrix $\hat{\mathbf{X}} = \sum_{i=1}^{m} \mathbf{X}_i$ ($1 \leq m \leq d$)) where d is the trace length, is defined as an approximation to the original tracing segment (either horizontal or vertical). In this application, $\mathbf{X}_i$ refers to the ith eigentriple and m has been considered as $m = 12$. Finally, the horizontal and vertical $\mathbf{X}$ are reconstructed and considered as the estimated PVDM after thresholding.

7.4.2 Blood Vessel Reconstruction and Linking the Vessel Segments

The initial estimated PVDM does not affect the accuracy of the tracking results because some false positives are often resolved by applying suitable post-processing techniques. Effectively, the estimated PVDM provides us with a set of curve fragments and their endpoints for tracking. If the size of curve segment is larger than or equal to three pixels, it is considered as the potential centreline of the vessels. Otherwise, the segment is considered to be noise and removed. The edge points are determined by estimating of the directions and half width of vessel R of each point of curve segments. Initially, R is predefined to eight and then updated. The cross-sectional profile of each vessel point g is obtained along a scan-line perpendicular to the current direction $\mathbf{g} = [g_1, g_2, ...g_W]^T$ (Figure 7.7 (a)). The width of the search window W is also adapted to the half width of lumen R, $W = 3R + 1$.

The *ith* element in this profile vector can be measured as:

$$\mathbf{g}[i] = I[x + (i - w - 1)\cos\theta, y - (i - w - 1)\sin\theta], \quad w = \frac{W-1}{2}. \quad (7.1)$$

Vessel level V is identified by the average intensity of the $2R + 1$ pixels around the current point. The remaining pixel values are averaged to identify a background level B as

$$V = \frac{1}{2R+1}\Sigma_{i=w-R+1}^{w+R+1}\mathbf{g}[i], \quad (7.2)$$

$$B = \frac{1}{2(w-R)}(\Sigma_{i=1}^{w-R}\mathbf{g}[i] + \Sigma_{i=w+R+2}^{2w+1}\mathbf{g}[i]). \quad (7.3)$$

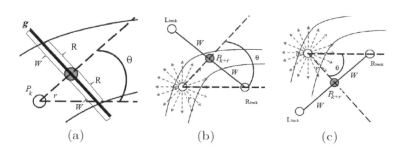

(a)   (b)   (c)

FIGURE 7.7: Geometric illustration of tracking process; (a) the process of updating the width of blood vessel, (b) and (c) the process for estimating the vessel pixels (arrows represent 16 searching directions).

The two edge points are determined by searching for the points whose levels are equal to or less than $(V + B)/2$ on both sides of the vessel point. Once the edge points are determined, the vessel half width R is updated. The search is repeated for all the curve fragments.

For each segment, the endpoints of the segment are used to estimate the next tracking points and connect other branches using the tracking approach. This approach extends the work of Vlachos and Dermatas for line tracking [26]. The set of next candidate tracking points, defined as C_{next}, are the 16 nearest neighbours of the current tracking point of P_k (Figures 7.7 (b) and (c)). The cross-sectional profile parameters V_p of the candidate pixels are denoted as:

$$
\begin{aligned}
V_p((x,y),(r,\theta_i)) \;=\;\; & I(x+(r\cos\theta_i-w\sin\theta_i), y+(r\sin\theta_i+w\cos\theta_i)) \\
+\;\; & I(x+(r\cos\theta_i+w\sin\theta_i), y+(r\sin\theta_i-w\cos\theta_i)) \\
-\;\; & 2I(x+(r\cos\theta_i), y+(r\sin\theta_i)),
\end{aligned}
\tag{7.4}
$$

where $(x,y) \in C_{next}$ and W is larger than the width of the blood vessel cross section. r is the distance between the current tracking point and the next candidate tracking point, for example $r = 12$ when the vessel width in the images is between seven to ten pixels. θ_i refers to the ith (out of 16) nearest neighbour. In Figure 7.7 P_k is the current tracking pixel and P_{k+r} is a candidate tracking pixel within C_{next}. R_{back} and L_{back} are the right and left candidate background pixels located $(W-1)/2$ pixels away from P_{k+r}. If P_{k+r} is a pixel on the blood vessel, the V_p is a large positive value as in Figure 7.7 (b). The parameter V_p may be closed to zero or a negative value when the candidate pixel is located in the background region Figure 7.7 (c). The direction of the vessel at P_k is defined by measuring the maximum positive parameter and a predefined threshold T:

$$
\theta_0 = \arg\max_{\theta_i}(V_p((x,y),(r,\theta_1))) : V_p((x,y),(r,\theta_i)) > T, 1 \le i \le 16). \tag{7.5}
$$

Then, update the tracking pixel to P_{k+r} which is located at:

$$
(x,y) = (x_k + r\cos\theta_0, y_k + r\sin\theta_0). \tag{7.6}
$$

The search continues for each vessel segment while the traced segments are recorded. The tracking method is terminated when all the endpoints have been processed. Each tracking procedure started from an endpoint will stop if one or more of the following conditions are satisfied.

1. The cross-sectional profile parameters V_p are less than a threshold T.

2. The new point is outside the image ROI.

3. A previously detected vessel intersects with the current one.

7.4.3 Post Processing

Since branching and crossover points could happen anywhere along a vessel, a back tracing mechanism is implemented to search for these points by moving along the vessel. This process is similar to finding the starting point, involving the following steps:

1. Determining a semi-circular region of small radius around a traced edge point.

2. Extracting the cross-sectional profile parameter of the pixel for this circle.

3. Determining the new starting point and storing it.

Each of these new starting points is checked to see if it has already been traced. The radius must be larger than the width of capillaries to make sure the semi-circular region includes both the vessel and non-vessel cross-section areas. On the other hand, too large a radius may cause many false positives. Here, the predefined radius has been set to 16 pixels.

7.4.4   Implementation and Results

The above method was tested on the DRIVE dataset (40 images) [12] and retinal images from different populations collected from Mongolia (1680 images), Kenya (9587 images), Botswana (500 images), Norway (840 images) and the UK (1000 images) [9].

Some of the images are from the diabetes patients found through diabetic retinopathy screening, while some are from pure population based studies where most subjects are expected to be normal. Figures 7.8 (a) and (b) show two examples from the Norway and Mongolia datasets, which have different qualities, contrasts and retinal pigment backgrounds. Figures 7.8 (c) and (d) show the extracted vessel networks in these two retinal images. The proposed method can extract some fine vessels and prevent some false positive rates, for example, the false vessel segmentation around the optic disc.

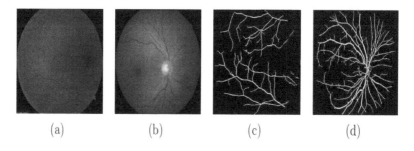

(a)   (b)   (c)   (d)

FIGURE 7.8: (a) and (b) fundus image from Norway and Mongolia datasets, (c) and (d) the corresponding two different vessel segmentation results.

In this work, the performance of the proposed approach has been evaluated based on accuracy, sensitivity, specificity, and efficiency. The accuracy is denoted as the ratio of the number of correctly classified vessel and non-vessel pixels to the total number of pixels in the image field of view. The sensitivity

(SN) is measured by the proportion of true positives correctly identified and specificity (SP) refers to the proportion of detected true negatives.

Two hundred images have been selected randomly from different datasets. A ground truth for these images has been established by employing trained observers. These manual annotations have been used to evaluate the performance of the method.

Table 7.1 illustrates the performance of the segmentation algorithm for various datasets. It is observed that the average accuracy for all the datasets is approximately 0.9435, and the average SN and SP are 0.7235 and 0.9761 respectively. The performance of the proposed technique is compared with those of two published algorithms for DRIVE dataset in Table 7.2. In addition, the running time per image (935 × 935) is within one minute. The results indicate a good outcome in terms of accuracy and reliability of the approach for processing of the retinal images across various populations.

TABLE 7.1:   Performance measure of the above method on the different datasets.

Database	Accuracy	Sensitivity	Specificity
Mongolia	0.9415	07210	0.9781
Kenya	0.9530	0.7782	0.9786
Botswana	0.9368	0.7285	0.9818
Norway	0.9350	0.7097	0.9808
UK	0.9518	0.7520	0.9782
Average	0.94335	0.7235	0.9761

TABLE 7.2:   Performance comparison of vessel segmentation methods (DRIVE images).

Methods	Accuracy	Sensitivity	Specificity
Staal et al.'s	0.9441	NA	NA
Vlachos and Dermatas's	0.9290	0.7470	0.9550
Proposed	0.9432	0.7185	0.9718

7.5   Concluding Remarks

The importance and significance of processing of ocular fundus images, particularly digital diabetic retinopathy images, have been emphasised and the role of SSA in this area described.

The change point detection property of SSA has been well exploited in this work for vessel extraction by integrating singular spectrum analysis and local vessel tracing. The results of this work have been presented and the approach validated by using the retinal images from various ethnic groups. It has been also demonstrated the accuracy and significant reduction on false vessel segmentations even for small vessels (capillaries).

Bibliography

[1] Abrámoff, M. D., Reinhardt, J. M., Russell, S. R., Folk, J. C., Mahajan, V. B., Niemeijer, M., and Quellec, G. (2010). Automated early detection of diabetic retinopathy. *Ophthalmology*, **117**(6), 1147–1154.

[2] WHO. (2012). *Diabetes programme*, last accessed on 23rd October 2013.

[3] Committee, U. N. S. (2012). *NHs diabetic eye screening programme*. Last accessed on 23rd October 2013.

[4] Rull, D. G. (2011). *Diseases and different ethnic groups*. Last accessed on 18th November 2013.

[5] Simmons, D., Clover, G., and Hope, C. (2007). Ethnic differences in diabetic retinopathy. *Diabetic Medicine*, **24**(10), 1093–1098.

[6] Sivaprasad, S., Gupta, B., Crosby-Nwaobi, R., and Evans, J. (2012). Prevalence of diabetic retinopathy in various ethnic groups: a worldwide perspective. *Survey of ophthalmology*, **57**(4), 347–370.

[7] Poe, M. (2013). Ophthalmic photography. Last accessed on 18th November 2013.

[8] Peek. (2013). Portable eye examination kit. Last accessed on 23rd October 2013.

[9] Wang, S., Tang, L., Hu, Y., Sanei, S., Alturk, L., Saleh, G. and Peto, T. (2014). Retinal vessel extraction by integrating singular spectrum analysis and vessel tracing. *Submitted to MICCAI 2014*, Boston, USA.

[10] Yu, H., Barriga, S., Agurto, C., Echegaray, S., Pattichis, M., Zamora, G., Bauman, W. and Soliz, P. (2011). Fast localization of optic disc and fovea in retinal images for eye disease screening. In *SPIE Medical Imaging*, 796317. International Society for Optics and Photonics.

[11] Winder, R. J., Morrow, P. J., McRitchie, I. N., Bailie, J. R., and Hart, P. M. (2009). Algorithms for digital image processing in diabetic retinopathy. *Computerized Medical Imaging and Graphics*, **33**(8), 608–622.

[12] Staal, J., Abrámoff, M. D., Niemeijer, M., Viergever, M. A., van Ginneken, B., Ridge-based vessel segmentation in color images of the retina. *Medical Imaging, IEEE Transactions*, **23**(4), 501–509.

[13] Rigamonti, R., and Lepetit, V. (2012). Accurate and efficient linear structure segmentation by leveraging ad hoc features with learned-lters. In *Medical Image Computing and Computer-Assisted Intervention MICCAI*, 189–197. Springer Berlin Heidelberg.

[14] Becker, C., Rigamonti, R., Lepetit, V., and Fua, P. (2013). Supervised feature learning for curvilinear structure segmentation. In *Medical Image Computing and Computer-Assisted Intervention MICCAI*, 526–533. Springer Berlin Heidelberg.

[15] Chaudhuri, S., Chatterjee, S., Katz, N., Nelson, M., Goldbaum, M. (1989). Detection of blood vessels in retinal images using two-dimensional matched Filters. *IEEE transactions on medical imaging*, **8**(3), 263–269.

[16] Zhang, B., Zhang, L., Zhang, L., Karray, F. (2010). Retinal vessel extraction by matched filter with first-order derivative of Gaussian. *Computers in biology and medicine*, **40**(4), 438–445.

[17] Chutatape, O., Zheng, L., and Krishnan, S. M. (1998). Retinal blood vessel detection and tracking by matched Gaussian and Kalman Filters. In *Engineering in Medicine and Biology Society. Proceedings of the 20th Annual International Conference of the IEEE*, **6**, 3144–3149.

[18] Can, A., Shen, H., Turner, J. N., Tanenbaum, H. L., and Roysam, B. (1999). Rapid automated tracing and feature extraction from retinal fundus images using direct exploratory algorithms. *Information Technology in Biomedicine, IEEE Transactions*, **3**(2), 125–138.

[19] Wang, L., Kallem, V., Bansal, M., Eledath, J., Sawhney, H., Karp, K., Stone, R. A. (2013). Interactive Retinal Vessel Extraction by Integrating Vessel Tracing and Graph Search. In *Medical Image Computing and Computer-Assisted Intervention MICCAI*, 567–574. Springer Berlin Heidelberg.

[20] Tang, H. L., Goh, J., Peto, T., Ling, B. W. K., Hu, Y., Wang, S., and Saleh, G. M. (2013). The reading of components of diabetic retinopathy: An evolutionary approach for filtering normal digital fundus imaging in screening and population based studies. *PloS one*, **8**(7), e66730.

[21] Sinthanayothin, C., Boyce, J. F., Williamson, T. H., Cook, H. L., Mensah, E., Lal, S., and Usher, D. (2002). Automated detection of diabetic retinopathy on digital fundus images. *Diabetic medicine*, **19**(2), 105–112.

[22] Walter, T., Klein, J. C., Massin, P., and Erginay, A. (2002). A contribution of image processing to the diagnosis of diabetic retinopathy-detection of exudates in color fundus images of the human retina. *Medical Imaging, IEEE Transactions*, **21**(10), 1236–1243.

[23] Gardner, G. G., Keating, D., Williamson, T. H., and Elliott, A. T. (1996). Automatic detection of diabetic retinopathy using an artificial neural network: a screening tool. *British Journal of Ophthalmology*, **80**(11), 940–944.

[24] Fleming, A. D., Goatman, K. A., Philip, S., Olson, J. A., and Sharp, P. F. (2007). Automatic detection of retinal anatomy to assist diabetic retinopathy screening. *Physics in Medicine and Biology*, **52**(2), 331.

[25] Golyandina, N., Nekrutkin, V., Zhigljavsky, A. A. (2001). *Analysis of time series structure: SSA and related techniques*. CRC Press.

[26] Vlachos, M., and Dermatas, E. (2010). Multi-scale retinal vessel segmentation using line tracking. *Computerized Medical Imaging and Graphics*, **34**(3), 213–227.

Chapter 8

Prediction of Medical and Physiological Trends

Healthcare improvement, marketing pharmaceutical products, prevention of epidemic diseases, and effective planning for future clinical services are very dependent on how the previous and current trends in the states of the above elements can be described. On the other hand, the adaptive systems for rehabilitation and accident avoidance, including autonomous systems, can gain significantly in places where the future state of the system is accurately estimated. The aim of SSA is twofold:

- To make a decomposition of the original series into a sum of a small number of independent and interpretable components such as a slowly varying trend, oscillatory components and structureless noise;

- To reconstruct the decomposed series so as to make predictions without the noise component.

As for economical and sociological information, there are numerous trends in clinical statistics which can be effectively exploited in prediction and identification of variables such as:

- Use of pharmaceutical products

- Propagation of epidemic diseases

- Transmission of drug resistant bacteria

- Progress of benign and malignant tumours

- Cure progress for chronic diseases and wounds

- Lifetime prediction of biological specimens

and many other cases. Therefore, establishing a good predictor for such applications may become vital for both monitoring and prevention.

This chapter introduces very popular methods in time series prediction and then looks back into the prediction capability of SSA, as the main focus of the chapter.

8.1   Conventional Approaches for Time Series Prediction

Prediction of time series involves estimation of future samples of the data from the existing samples of the sequence. This is often called multi-step ahead prediction.

A series in hand usually includes a smooth background trend, some cyclic variations, anomalies, and noise. The prediction methods try to retain the main trend and the cyclic information while rejecting the anomalies, transients, and noise.

There are three major difficulties for accurate forecasting of a time series. Firstly, the patterns of most time series are nonstationary, i.e. there is no single fixed model that can describe the entire time series. Secondly, it is hard to make a compromise between the long-term trend and short-term sideways movements during the training process. In other words, an efficient system must be able to adjust its sensitivity in time. Thirdly, it is usually difficult to determine the usefulness of information. Misleading information must be identified and eliminated.

In the past, a number of approaches have been introduced for the above purpose. Most of these methods fall within the category of *data regression*. They involve modelling the mechanism through which the data has been generated. This model is later used for the prediction of future data samples.

In the following subsections we briefly review the regression techniques used for time series prediction.

8.1.1   Multiple Linear Regression

Multiple linear regression (MLR) defines a linear connection between the past and future samples. Autoregressive (AR) and multivariate autoregressive (MVAR) are the popular MLR methods. AR relies on single channel measurements and MVAR estimates a future sample for one channel using the previous samples of other measurements/time series.

Some biomedical applications of AR modelling have been explored in [2].

The main objective of using prediction methods is to find a set of model parameters which best describes the signal generation system. Such models generally require a noise type input. In AR modelling of signals each sample of a single channel measurement is defined to be linearly related to a number of its previous samples, i.e.

$$y(n) = -\sum_{k=1}^{p} a_k y(n-k) + x(n), \tag{8.1}$$

where a_k, $k = 1, 2, \ldots, p$, are the linear prediction coefficients of the AR model, n denotes the discrete sample time normalised to unity, and $x(n)$ is the input noise.

In some applications, an autoregressive moving average (ARMA) method is used instead. ARMA is a linear predictive model for which each new sample is estimated based on a number of its previous input and output sample values, i.e.

$$y(n) = \sum_{k=1}^{p} a_k y(n-k) + \sum_{k=0}^{q} b_k x(n-k), \tag{8.2}$$

where b_k, $k = 1, 2, \ldots, q$, are the moving average coefficients. The parameters p and q are the model orders. The Akaike criterion can be used to determine the order of the appropriate model of a measurement signal by minimizing the following equation [3] with respect to the model order of an ARMA model:

$$AIC(i,j) = N \ln(\sigma_{ij}^2 + 2(i+j)), \tag{8.3}$$

where i and j represent respectively, the assumed AR and MA model prediction orders, N is the number of signal samples, and $2\,\sigma_{ij}^2$ is the noise power of the ARMA model at the ith and jth stages. Later in this chapter we will see how the model parameters are estimated either directly or by employing some iterative optimisation techniques.

In an MVAR approach a multichannel scheme is considered. Therefore, each signal sample is defined versus both its previous samples and the previous samples of the signals from other channels, i.e. for channel i we have:

$$y(n) = -\sum_{j=1}^{m}\sum_{k=1}^{p} a_{jk} y_j(n-k) + x_i(n) \tag{8.4}$$

where m represents the number of channels and $x_i(n)$ represents the input noise to channel i. Similarly, the model parameters can be calculated iteratively in order to minimise the error between the actual and predicted values [4].

There are numerous applications for linear models. Different algorithms have been developed and proposed to find the model coefficients efficiently. In the maximum likelihood estimation (MLE) method [4]–[6] the likelihood function is maximised over the system parameters formulated from the assumed real, Gaussian distributed, and sufficiently long input signals of approximately 2500–5000 samples.

Using AIC the gradient of the squared error is minimised using the Newton-Raphson approach applied to the resultant nonlinear equations [6, 10]. This is considered as an approximation to the MLE approach. In the Durbin method [8] the Yule–Walker equations, which relate the model coefficients to the auto-correlation of the signals, are iteratively solved. The approach and the results are equivalent to those using a least-squared-based scheme [9]. The MVAR co-efficients are often calculated using the Levinson-Wiggines-Robinson (LWR) algorithm [10]. These parameters are then used to predict the new time series samples.

Any nonstationarity, wrong choice of the number of parameters (the order), and insufficient data length can impose severe error on both the estimation and the prediction processes.

8.1.2   Recurrent Neural Networks

There are a number of hybrid approaches which try to take advantage of the accuracy of statistical models and the generality of neural network approaches. Combination of autoregressive integrated moving average (ARIMA) and feed–forward neural networks (NNs) is an example.

ARIMA is an extension of ARMA and is generally referred to as an ARIMA(p, d, q) model. The parameters p, d, and q are non-negative integers that refer to the order of the AR, integrated, and moving average (MA) parts of the model respectively. ARIMA models form an important part of the Box–Jenkins approach to time-series modelling [7]. Each parameter then refers to 'AR', 'I' or 'MA' from the acronym describing the model. For example, ARIMA$(0,1,0)$ is I(1), and ARIMA$(0,0,1)$ is MA(1).

NNs maybe divided into two types of feed forward and recurrent NNs. A feed forward neural network (FFNN) is a type of NN where all of its connections have the same direction. This composition could be efficient in predicting many different time series. However, in some other time series, as expected, finding the proper value for the statistical part of the composition is a difficult task and incorrect values could affect the accuracy of prediction.

Recurrent NN (RNN), on the other hand, is a class of NNs where the connections between its layers can be two-sided, i.e. backward and forward, so, the connections between neurons form a directed cycle. Therefore, the recurrent networks can approximate any function by learning from both the inputs and the outputs.

Figure 8.1 shows a typical three-layer recurrent NN with four inputs, two outputs, and two neurons in its hidden layer.

Various optimisation methods may be used to find the link weights for the RNN. Giles et al. [9] used an RNN for prediction of a noisy time series for financial forecasting.

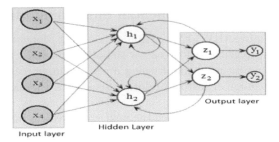

FIGURE 8.1: A typical three-layer recurrent NN with four inputs, two outputs, and two neurons in its hidden layer.

8.1.3 Hidden Markov Model

A hidden Markov model (HMM) is a finite state machine used for modelling nonlinear systems. There are three problems associated with HMM;

- given an HMM model, how the probability of detecting a particular sequence can be estimated,

- given the model and the sequence, what the state transitions are; often viterbi algorithm together with backward tracing is used at this stage; and

- given the number of states and the training sequence, how the model parameters can be estimated. In time series prediction using HMM, the model is built up using the known part of the sequence and it is used to predict the future states and the associated samples.

The HMM may have a Gaussian mixture at each state. Gaussian mixtures have desirable features as they can model series that do not fall into the single Gaussian distribution. The parameters of the HMM are updated after each iteration often by employing an add-drop Expectation-Maximization (EM) algorithm.

Shi and Weigend [11] used HMM for financial forecasting. The new state of the system is estimated using a fully connected trellis by Viterbi algorithm. Nobakht et al. [12] and Gupta et al. [1] also presented their HMM algorithms for prediction of time series.

In an interesting application [16] human movement has been predicted using an HMM. A next move of a person within an office environment has been predicted by this model. Both order 1 and order R HMMs have been implemented for this scenario. They showed that an order 1 HMM often outperforms order R HMM and various order Markov and NN models.

Dual HMMs [17] and Coupled HMMs [18] have also been defined in the literature in order to benefit from the correlation/dependence of two or more sequences in the estimation of their new states.

8.1.4   Holt–Winters Exponential Smoothing Model

The Holt–Winters (HW) Exponential Smoothing Model offers an alternative methodology to generate forecasts with both trend and seasonal variations through predicting future demand using a set of simple recursions that rely on a weighted average of historical data values, with the more recent values carrying more weight.

In the so–called Holt's approach the next step involves introducing a term to take into account the possibility of a series as a trend. In single exponential smoothing, the forecast function is simply the latest estimate of the level. If a slope component is now added which itself is updated by exponential smoothing, the trend can be taken into account.

For a series $y_1, y_2, \ldots, y_n$, the forecast function, which gives an estimate of the series h steps ahead can be written as:

$$\hat{y}_{n+h|n} = m_n + hb_n \quad l = 1, 2, \ldots, \tag{8.5}$$

where m_n is the current level and b_n is the current slope. Therefore, the one step ahead prediction is simply given by:

$$\hat{y}_{t|t-1} = m_{t-1} + b_{t-1}. \tag{8.6}$$

Since there are now two terms to the exponential smoothing, two separate smoothing constants as α_0 for the level and α_1 for the slope are used. As in single exponential smoothing, the updated estimate of the level m_t is a linear combination of $\hat{y}_{t|t-1}$ and y_t :

$$m_t = \alpha_0 y_t + (1 - \alpha_0)(m_{t-1} + b_{t-1}) \quad 0 < \alpha_0 < 1. \tag{8.7}$$

This provides the level at time, t. Since the level at time $t - 1$ is already known, it is possible to update the estimate of the slope:

$$b_t = \alpha_1(m_t - m_{t-1}) + (1 - \alpha_1)b_{t-1} \quad 0 < \alpha_1 < 1. \tag{8.8}$$

These equations can also be written in the appropriate error correction form:

$$m_t = m_{t-1} + b_{t-1} + \alpha_0 \varepsilon_t, \tag{8.9}$$

$$b_t = b_{t-1} + \alpha_0 \alpha_1 \varepsilon_t.$$

Holt's algorithm requires the initial values for m_t and b_t to be applied to the input, and estimates of the values for α_0 and α_1. Typical values are $0.02 < \alpha_0, \alpha_1 < 0.2$, but they can be estimated by minimising the sum of squared errors as in single exponential smoothing. Also, it is often found that $m_1 = y_1$, and $b_1 = y_2 - y_1$ are reasonable starting values.

Forecasting: Holt's method can be extended to deal with time series which contain both trend and seasonal variations. The HW method has two versions, additive and multiplicative, the use of which depends on the characteristics of the particular time series. The general forecast function for the multiplicative HW method is:

$$\hat{y}_{n+h|n} = (m_n + hb_n)\, c_{n-s+h} \quad l = 1, 2, \ldots, \tag{8.10}$$

where m_n is the component of level, b_n is the component of the slope, and c_{n-s+h} is the relevant seasonal component, with s signifying the seasonal period (e.g. 4 for quarterly data and 12 for monthly data.).

In addition to α_0 and α_1, here a third constant $0 < \alpha_2 < 1$, is added as the smoothing constant for the seasonal factor. The updating equations are therefore:

$$m_t = \alpha_0 \frac{y_t}{c_{t-s}} + (1 - \alpha_0)(m_{t-1} + b_{t-1}),$$

$$b_t = \alpha_1(m_t - m_{t-1}) + (1 - \alpha_1)b_{t-1}, \tag{8.11}$$

$$c_t = \alpha_2 \frac{y_t}{m_t} + (1 - \alpha_2)c_{t-s}.$$

For the additive version of HW the seasonal factor is simply added, as opposed to being multiplied, into the one step ahead forecast function, thus:

$$\hat{y}_{n+h|n} = m_n + b_n + c_{n-s+h}, \tag{8.12}$$

and the level and seasonal updating equations involve differences as opposed to ratios:

$$m_t = \alpha_0(y_t - c_{t-s}) + (1 - \alpha_0)(m_{t-1} + b_{t-1}), \tag{8.13}$$

$$c_t = \alpha_2(y_t - m_t) + (1 - \alpha_2)c_{t-s},$$

whereas the slope component, b_t, remains unchanged.

There are however, simple practical examples on how the HW method works; one is shown in the following simple example. To perform a monthly sales forecast, the HW method requires one to estimate up to three components of a forecasting equation namely, the current underlying level of sales, the current trend in our sales, and the seasonal index for the month we are forecasting. Given these parameters as current level = 500, our current trend = 5, and seasonal index = 1.2 for March. The forecast for the level in March will be:

$$
\begin{aligned}
\textit{The level forecast} \ &= \ [\textit{level} + 2 \times \textit{trend}] \times \textit{Seasonal index} \\
&= \ [500 + 2 \times 10] \times 1.2
\end{aligned}
$$

As soon as a new sales figure arrives, HW updates its estimates of the level, trend, and seasonal index for that month.

Many companies use the HW method to produce short-term demand forecasts when their sales data contain a trend and a seasonal pattern. The HW model has been used to separately forecast the distinct types of demand. The specific time series they investigated was an aggregate of two separate series recording emergency and non-emergency calls. Each model included terms to account for average, trend, seasonal and day-of-week effects.

8.2   SSA Application for Prediction

SSA has become an emerging technique in time series and event prediction. Application of this method for financial forecasting has been well established. One example can be seen in [19]. Application of SSA for prediction, however, has not been limited to financial forecasting. In [20] the authors describe the connections between time series analysis and nonlinear dynamics and present some of the novel methods for spectral analysis. The various steps, as well as the advantages and disadvantages of these methods, are illustrated by their application to an important climatic time series, the *Southern Oscillation Index*, which captures major features of inter-annual climate variability and is used extensively in its prediction. SSA performance has been evaluated when applied to regional and global sea surface temperature using a number of datasets.

In another research [26] SSA combined with NN has been used for prediction of daily rainfall. Through experiments they verified that the role of SSA in decomposition of the signals before applying the NN improves the prediction outcome significantly.

8.3   How Is SSA Used in Prediction?

SSA-based forecasting is performed through application of linear recurrent formulae (LRF) or equations. The class of series governed by LRF is rather wide; it contains harmonics, polynomials and exponential series and is closed under term-by-term addition and multiplication. An infinite series is governed by some LRF if and only if it can be represented as a linear combination of products of exponential, polynomial and harmonic series [22, 23].

We recall that, the single most important part of embedding is the selection of window length L, an integer such that $2 < L < N$ (N is the series length).

This selection should always be large enough to permit reasonable separability. It should not be greater the $N/2$ for optimum results. The vectors X_i, called the lagged vectors or L lagged vectors (to emphasise their dimension) form the K columns of the trajectory matrix $\mathbf{X}$.

If the rank of the trajectory matrix is smaller than the window length ($r < L$), the signal satisfies the LRF. For such signals, SSA could be applied as the forecasting algorithm. In usual practices, the reconstruction stage of SSA aims to smooth the original data by removing the eigentriples corresponding to noise. Then, a recurrent forecasting procedure is applied to the reconstructed signal [22].

Consider $\breve{q} \in C^{L-1}$ consisting of the last $L-1$ elements of the eigenvector $\mathbf{q}$ and $\nu^2 = \eta_1^2 + \ldots \eta_r^2$ in which η_i is the last element of the corresponding eigenvector. The vector $\delta = (\delta_1, \ldots, \delta_{\mathbf{L}-\mathbf{1}})\mathbf{T}$ is then defined as:

$$\delta = \frac{1}{1 - \nu^2} \sum_{i=1}^{r} \eta_i \breve{q}_i, \tag{8.14}$$

where i refers to index of the eigenvector. At this stage, the following equation is used to forecast h steps ahead using δ and the reconstructed signal $\hat{x}(n)$:

$$f(x) = \begin{cases} x(n) & n = 1, \ldots, N \\ \sum_{j=1}^{L-1} \delta_j \hat{x}(n-j) & n = N+1, \ldots, N+h \end{cases} \tag{8.15}$$

Note that Equation (8.15) caters to real-valued signals. To adapt such a procedure for complex data, prediction is calculated from the real and imaginary parts of the eigenvectors separately. This is because η_i represents the steering angle between a realvalued eigenvector and the x-axis (real axis). On the other hand, for complex-valued data, the angle corresponds to the phase between real and imaginary components of the vector. This cross-information will be exploited in the form of augmented statistics (see next section), and therefore is not considered here explicitly in our prediction task.

Using the above simple and easy to implement approach the future samples of the fundamental trends can be approximated. In practice any a priori about signal statistical properties, frequency band or even components' source locations may be used to enhance the performance of the prediction algorithm. For example, in [23] the spatial information has been exploited to improve the prediction. This has been performed by introducing an inverse distance weighting technique [24, 25] as a means of incorporating spatial information. In doing that, an additional explanatory variable by taking spatially weighted averages has been included.

One way to improve the results is to consider any cyclic component as a separate component and predict it separately. It is known by now that SSA is able to decompose daily rainfall series into several additive components that typically can be interpreted as trend components (which may not exist), various oscillatory components, and noise components. Therefore, this approach requires decomposition of the signal into its main and cyclic trends before going to the prediction stage. The LRF property of the SSA implies that the prediction of the overall trend is equivalent to the sum of the predicted values of its constituent components. An important advantage of such an approach is that the SSA parameters can be defined for the individual components separately.

To clarify the concept, look at a synthetic 2500 sample trend depicted in Figure 8.2.

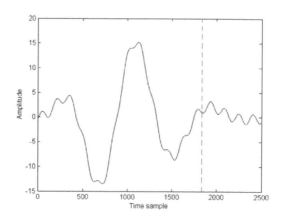

FIGURE 8.2: A synthetic trend comprising a smooth trend and a cyclic trend. The dashed line shows the start point of prediction.

In practice there might be various noise added to this trend too. If we consider the trend up to sample 1800 and apply the basic SSA prediction described above, using a window length $L = 400$, to predict up to sample 2200, the trend in Figure 8.3 will be achieved.

It is clear that the cyclic part of the trend has not been well reconstructed. Now if the original trend is decomposed primarily into two components (using SSA, as described in Chapter 2) as in Figure 8.4 and the prediction is performed on individual components (using $L = 500$ for the smooth component and $L = 200$ for the cyclic component) to estimate the samples from 1800 afterward, the trend looks like that in Figure 8.5.

For well-structured time series often recurrent NN outperforms other methods including basic SSA. Other methods such as AR and HMM often fail to predict long term trends. The AR method normally requires a low prediction order, p, in order to avoid noise. Therefore, its prediction capability significantly deteriorates beyond p samples. For the HMM, increasing the number of states increases the computational complexity exponentially, and to cope with a reasonable prediction length a very large memory size is required.

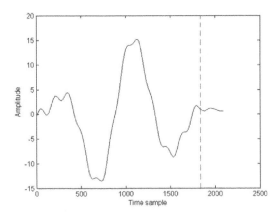

FIGURE 8.3: The prediction of the samples 1800–2500 of the trend in Figure 8.2.

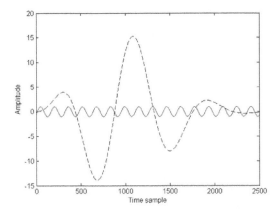

FIGURE 8.4: The constituent components of the trend in Figure 8.2.

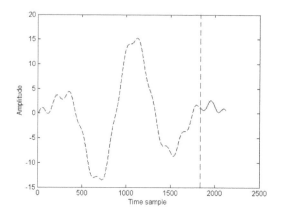

FIGURE 8.5: Prediction results when the two series components are treated separately.

However, as explained above, various a priories can be easily incorporated into the SSA-based prediction algorithm to enhance the outcome.

8.4 Application of SSA-Based Prediction to Real Biomedical Trends

8.4.1 Prediction of Pharmaceutical Product

In an application of the SSA to prediction of clinical information the following time series from a pharmaceutical company has been considered. This trend has been sampled over 8 years, once a week regularly, starting from January 1. The trend expresses antihistamine consumption within one of the European states. The trend is depicted in Figure 8.6.

It is clearly seen that there is a cyclic trend, which is likely to be due to the springtime when there is more demand for anti-allergy drugs.

The above two methods i.e. by directly applying the SSA to the trend and by first decomposing the data (using another SSA) into the baseline and cyclic trends and then applying SSA, have been applied to this trend. The results can be seen in Figure 8.7, denoted respectively by dashed line ($L = 250$) and blue bold line ($L = 100$).

A more accurate prediction of the market situation for many drugs provides a good indicator for health monitoring and control.

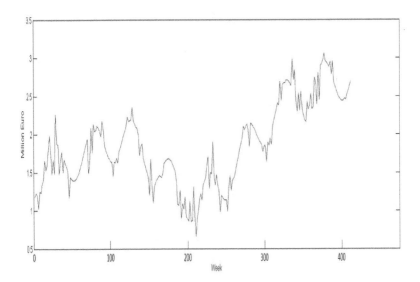

FIGURE 8.6: The trend for Antihistamine consumption.

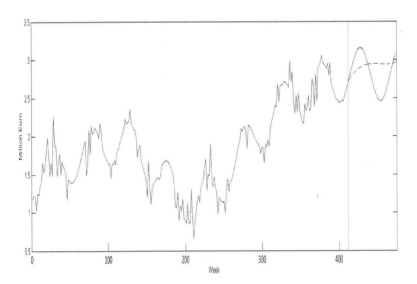

FIGURE 8.7: Predicted samples (from 416 to 475 weeks) using basic SSA (dashed line) and decomposition + prediction using SSA (bold line).

8.4.2 Predicting Ambulance Demand

Traditionally, Emergency Medical Services (EMS) system managers utilise historical data to calculate their anticipated call volumes via a method of prediction known as demand pattern analysis.

Brown et al. [28] analysed three variations of the method: average peak demand, smoothed average peak demand and 90th percentile rank, to predict demand for each hour of each day for 52 weeks based on known values for an initial 20 week period. Average peak demand makes predictions for each hour based upon the average of the highest number of calls recorded for that hour in the first 10 and last 10 weeks of the initial 20 week period. On the other hand, smoothed average peak demand takes a weighted average of the average peak prediction for the hour in question along with the preceding and following hours. 90th percentile rank distinctly considers the demand for each of the 20 weeks in the initial period, and takes the value at the 90th percentile from the rank list.

The authors concluded that the demand pattern analysis generally either accurately estimated or overestimated call volume, making it a reasonable predictor for ambulance staffing patterns [28]. Particularly, they discovered that 90th percentile rank accurately predicted call volume more consistently than the other methods. However, although this was found to be the best of the three methods, it was shown to accurately predict call volumes (± 1 call) only 19% of the time. It underestimated the demand 7% of the time, and overestimated demand the remaining 74% of the time. It is at the discretion of individual communities to determine if these over- and under-estimation rates are acceptable for their situation.

Although the demand pattern analyses involve some stochastic variation of the inputs, they do not necessarily take into account seasonal variations and other stochastic effects that might arise. This causes a great deal of information to be lost through using summary measures to inform the forecasts in place of the data itself [29]. Furthermore, the predictions generated by the demand pattern analysis techniques are infrequently updated in practice, meaning the coverage requirements are often repeated on cyclical bases, and the operation can provide inaccurate forecasts if the peak demand values are extreme.

It is therefore imperative to develop appropriate estimates of future demand since excessively large estimates lead to over-staffing and unnecessarily high costs, while low estimates lead to under-staffing and slow response times. In recognition of the shortfalls of the elementary methods to produce accurate demand forecasts, numerous statistical models have been investigated and developed by operational researchers to provide advanced forecasts by simultaneously dealing with trend, seasonal fluctuations, and random error.

In the early 1970*s*, several researchers started to tackle the problem of ambulance demand prediction; attempting to explain the demand in small regions of the USA using a limited number of demographics, alongside land use and socio-economic variables as discussed in [30]–[32].

Most of the earliest models, based on multiple regression, were performed on incomplete datasets with outdated socioeconomic and population data; nevertheless they generated models capable of predicting total yearly demand to a high degree of accuracy due to the significant effects of sociodemographic characteristics.

Aldrich et al. [30] used 31 independent variables to predict per capita EMS demand in 157 areas of Los Angeles in addition to per capita demand for 6 different incident types (namely traffic accidents, other accidents, dry runs, cardiac cases, poison cases and other illnesses).

Their paper concluded that a linear model employing sociodemographic variables was capable of predicting total ambulance demand to a high degree of accuracy and found that areas with high densities of low-income families, non-whites, elderly people or children tended to use the public ambulance service more often than others. They also found that socioeconomic variables didn't need to be as powerful in the analysis as other variables, but housing density and land use had some effect.

Although Aldrich [30] noted in his work that the method described in his paper would be applicable to any region, he highlighted that the specific regression model would not.

Aldrichs work was extended by several authors such as Kvalseth and Deems [31] and Kamentzky et al. [32], who recognised that yearly demand for public ambulances appeared to be highly predictable within a certain city or country.

By simply adjusting the variables considered, they developed regression models capable of predicting demand variations for specific ambulance trusts.

In further work by Kamentzky et al. [32] he described the variation in demand in Southwestern Pennsylvania using only four independent variables selected using both stepwise and forced entry procedures. The final model for representing the total demand provided a very good fit to the data.

Some information such as age of the population was exploited in a later work by McConnel and Wilson [33]. They found that the pattern of utilisation of EMS services associated with age was tri-modal, with rates rising geometrically for individuals aged 65 and over, to the extent that the utilisation rate was 3.4 times higher for those aged 85 years and over compared to those aged 45–64.

Vile et al. [27] used SSA to predict the demand exerted upon the EMS using data provided by the Welsh Ambulance Service Trust (WAST). The aim of this approach is to explore new methods to produce accurate forecasts that can be subsequently embedded into current systems to optimise the resource allocation of vehicles and staff, and allow rapid response to potentially life-threatening emergencies. They claim that SSA produces superior longer-term forecasts (which are especially helpful for EMS planning), and comparable shorter-term forecasts to well established methods [26].

In an application of SSA for demand prediction, a 1-month SSA forecast beginning on 1st July is given in Figure 8.8. All 1,552 daily counts up to 30th June 2009 are used in the estimation of the SSA model shown, although only the within-sample data for the month of June 2009 is displayed for clarity. In addition to the root mean-square error (RMSE), visual inspection of Figure 8.8 shows that the HW method captures some element of the periodic nature of demand, but not the full variation of peaks and troughs throughout July. In contrast the SSA and ARIMA forecasts follow the true demand values reasonably closely, but of the two methods, the SSA forecast maintains the lowest RMSE when the rolling forecast is computed until December.

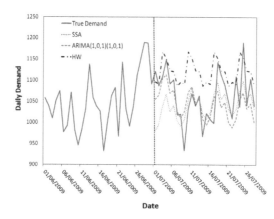

FIGURE 8.8 (See color insert.): 28-day forecasts commencing on 1st July 2009 [25].

The benefit of the SSA technique as seen in this application is not only in its ability to forecast but in its capability to recognise periodicities in the data and its flexibility and ease of implementation.

8.4.3   Progress of Alzheimer's Disease and Prediction of Critical State of the Patient

In March 2014, it was announced in *New Scientist Magazine* that the world's first blood test to predict Alzheimer's disease (AD) before symptoms occur had been developed. The test identifies 10 chemicals in the blood associated with the disease two to three years before symptoms start, but it might be able to predict Alzheimer's decades earlier. Globally, 35 million people suffer from Alzheimer's. AD is characterised by a toxic build up of amyloid and tau proteins in the brain, which destroys the neurons. Several blood tests can diagnose the disease, but until now, none has had the sensitivity to predict its onset. In parallel with blood tests, conventional cognitive and memory tests, such as Adenbrookes Cognitive Examination (ACE), Montreal cognitive assessment (MoCA), or Mini mental state examination (MMSE), are also performed on a regular basis to evaluate the symptoms and reject other possibilities such as mild cognitive impairment (MCI) [34] or other forms of dementia.

Based on this article, initially published in *Nature Medicine* [35], an analysis of the participants' blood highlighted ten metabolites that were depleted in those with mild cognitive impairment who went on to get Alzheimer's compared with those who did not. In subsequent trials, the researchers showed these chemicals could predict who would go on to get Alzheimer's within the next three years with up to 96% accuracy.

The suggested ten metabolites play a key role in supporting cell membranes, maintaining neurons or sustaining energy processes. It is hoped that once verified in a larger group, the test should provide a cheap and quick way of predicting Alzheimer's.

Mapstone says that it may even be able to predict the disease much earlier, because the brain changes associated with Alzheimer's begin many years before symptoms occur. These metabolic changes might occur 10 or 20 years earlier; that would give us a real head start on predicting the disease [35].

The researchers in George Town also analysed the full genome sequence of all of the participants in the study. That work has yet to be published, but Federoff believes that the gene changes are linked to the metabolite changes. So, by combining the information, they hope to provide a more complete description of the underlying pathology of the disease, and to know the function of all the affected genes, and, if done and they can intercept the changes, they might make good candidates for new drugs."

Progression of AD has also been studied through analysis of electroencephalograms (EEGs) aiming at replacing some EEG-based descriptors of AD for those achieved using cognitive tests such as ACE, MoCA, and MMSE. For example, a number of research articles on using EEG to assess the AD level such as [36] have been published.

In a collaborative work between the University of Surrey, UK, and the Psychology Department in the Czech Republic, the ACEs were undertaken from 28 AD patients. The examinations were taken regularly every three months over four years. In parallel with ACE, their EEG signals were also recorded and analysed using the connectivity patterns between the pair of electrodes with maximum connectivity in the most related frequency bands (These electrode pairs are F3-T6, F3-P4, T5-C3, P3-T5, P3-C3, C4-T5, C4-T5, C4-C3, C4-P3 within the theta band, C4-T5 within the alpha band, and T5-C3 and P3-C3 within the beta band.).

The ACE data were then interpolated/upsampled (1:10) and SSA was applied to predict the future state of the disease for all the patients for the next nearly two years. The trends and the prediction results are depicted in Figures 8.9 and 8.10 respectively for two of the patients I and II. From the trend and its prediction it is seen that although both trends are very close at the start, the trend for patient II progressed with a lower increasing slope. These results were noted. Further examinations by analysis of EEG and the estimation of connectivity, using multivariate autoregressive (MVAR) and directed transfer functions (DTF) methods [1] verified that patient II does not have AD, but suffers from MCI.

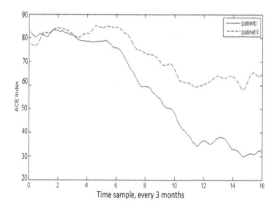

FIGURE 8.9: The ACE index trends over four years for the two patients, I, and II.

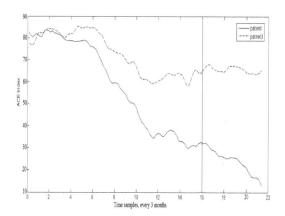

FIGURE 8.10: The predicted ACE index. The prediction starts after four years, i.e. after sample instant 16.

Figure 8.10 shows that for the MCI patient, i.e. patient I, (later diagnosed using other tests including EEG analysis) has a lower slope in deterioration of memory compared to a reasonably severe case of AD patients (i.e. patient II).

8.5 Concluding Remarks

This chapter points out the importance and significance of the prediction ability of SSA in monitoring health care, ambulatory cases, and the huge financial complexity involved in this area. The examples given in this chapter are indeed too few to be significant in such a demanding area of research. Both socioeconomical and patient monitoring aspects require more research to be undertaken and both need a robust, comprehensive, and inclusive approach in prediction of the future state of the system. SSA takes into account various underlying information within time series, such as rhythmicity, which is involved in many current healthcare problems.

Bibliography

[1] Sanei, S. (2013). *Adaptive Processing of Brain Signals*. Wiley.

[2] Akaike, H. (1974). A new look at the statistical model identification. *Automatic Control, IEEE Transactions*, **19**(6), 716–723.

[3] Steven, M. K. (1988). Modern spectral estimation: theory and application. *Signal Processing Series*.

[4] Gueguen, C., and Scharf, L. L. (1980). Exact maximum likelihood identification of ARMA models: a signal processing perspective. *Signal Processing Theory Applications*, M. Kunt and F. de Coulon, Eds., 759–769.

[5] Akay, M. (2001). *Biomedical Signal Processing*. Academic Press.

[6] Steven, M. K. (1988). Modern spectral estimation: theory and application. *Signal Processing Series*.

[7] Akaike, H. (1974). A new look at the statistical model identification. *Automatic Control, IEEE Transactions*, **19**(6), 716–723.

[8] Durbin, J. (1959). Efficient estimation of parameters in moving average models. *Biometrika*, **46**, 306–316.

[9] Trench, W. F. (1964). An algorithm for the inversion of finite Toeplitz matrices. *Journal of the Society for Industrial & Applied Mathematics*, **12**(3), 515–522.

[10] Morf, M., Vieira, A., Lee, D. T., and Kailath, T. (1978). Recursive multi-channel maximum entropy spectral estimation. *IEEE Transactions Geoscience Electronics*, **16**(2), 85–94.

[11] Box, G., and Jenkins, G. (1970). *Time series analysis: Forecasting and control*. San Francisco.

[12] Giles, C. L., Lawrence, S., and Tsoi, A. C. (2001). Noisy time series prediction using a recurrent neural network and grammatical inference. *Machine Learning*, **44**, 161–183.

[13] Shi, S., and Weigend, A. S. (1997, March). Taking time seriously: Hidden Markov experts applied to financial engineering. In *Computational Intelligence for Financial Engineering (CIFEr), Proceedings of the IEEE/IAFE*, 244–252.

[14] Nobakht, B., Joseph, C. E., and Loni, B. (2012). Stock market analysis and prediction using hidden markov models. In *Student Conference on Engineerig and Systems (SCES)*, 1–4.

[15] Gupta, A., and Dhingra, B. (2012). Stock market prediction using Hidden Markov Models. *Proceedings of Int. Conference on Engineering and Systems*, 1–4, India.

[16] Gellert, A., and Vintan, L. (2006). Person movement prediction using hidden Markov models. *Studies in Informatics and control*, **15**(1), 17.

[17] Dasguptaa, N., Runkleb, P., Couchmana, L., and Carina, L. (2001). Dual hidden Markov model for characterizing wavelet coefficients from multi-aspect scattering data. *Journal of Signal Processing*, **81**(6), 1303–1316.

[18] Brand, M., Oliver, N., and Pentland, A. (1997). Coupled hidden Markov models for complex action recognition. *Proceedings of IEEE Conf. on Computer Vision and Pattern Recognition*, 994–999, San Juan.

[19] Hassani, H., Heravi, S., and Zhigljavsky, A. (2013). Forecasting UK industrial production with multivariate singular spectrum analysis. *Journal of Forecasting*, **32**(5), 395–408.

[20] Ghil, M., Allen, M. R., Dettinger, M. D., Ide, K., Kondrashov, D., Mann, M. E., Robertson, A. W., Saunders, A., Tian, Y., Varadi, F., and Yiou, P. (2002). Advanced spectral methods for climatic time series. *Reviews of Geophysics*, **40**(1), 3.1-3.41.

[21] Vautard, R., and Ghil, M. (1989). Singular spectrum analysis in nonlinear dynamics, with applications to paleoclimatic time series. *Physica D: Nonlinear Phenomena*, **35**(3), 395–424.

[22] Golyandina, N., Nekrutkin, V., and Zhigljavsky, A. A. (2001). *Analysis of Time Series Structure: SSA and Related Techniques*. Chapman and Hall; CRC, New York, London.

[23] Awichi, R. O., and Müller, W. G. (2013). Improving SSA predictions by inverse distance weighting. *REVSTATC Statistical Journal*, **11**(1), 105–119.

[24] Bakkali, S., and Amrani, M. (2008). About the use of spatial interpolation methods to denoising Moroccan resistivity data phosphate "Disturbances" map. *Acta Montanistica Slovaca*, **13**(2), 216–222.

[25] Shepard, D. (1968). A two-dimensional interpolation function for irregularly-spaced data. In *Proceedings of the 1968 23rd ACM national conference*, 517–524. ACM, New York, NY, USA.

[26] Chau, K., and Wu, C. (2010). A hybrid model coupled with singular spectrum analysis for daily rainfall prediction. *Journal of Hydroinformatics*, **12**(4), 458–473.

[27] Vile, J. L., Gillard, J. W., Harper, P. R., and Knight, V. A. (2012). Predicting ambulance demand using singular spectrum analysis. *Journal of the Operational Research Society*, **63**(11), 1556-1565.

[28] Brown, L. H., Lerner, E. B., Larmon, B., LeGassick, T., and Taigman, M. (2007). Are EMS call volume predictions based on demand pattern analysis accurate? *Prehospital Emergency Care*, **11**(2), 199–203.

[29] Gillard, J., and Knight, V. (2013). Using Singular Spectrum Analysis to obtain staffing level requirements in emergency units. *Journal of the Operational Research Society*, **65**(5), 735–746.

[30] Aldrich, C. A., Hisserich, J. C., and Lave, L. B. (1971). An analysis of the demand for emergency ambulance service in an urban area. *American Journal of Public Health*, **61**(6), 1156–1169.

[31] Kvålseth, T. O., and Deems, J. M. (1979). Statistical models of the demand for emergency medical services in an urban area. *American Journal of Public Health*, **69**(3), 250–255.

[32] Kamenetzky, R. D., Shuman, L. J., and Wolfe, H. (1982). Estimating need and demand for prehospital care. *Operations Research*, **30**(6), 1148–1167.

[33] McConnel, C. E., and Wilson, R. W. (1998). The demand for prehospital emergency services in an aging society. *Social Science & Medicine*, **46**(8), 1027–1031.

[34] Escudero, J., Sanei, S., Jarchi, D., Abásolo, D., and Hornero, R. (2011). Regional coherence evaluation in mild cognitive impairment and Alzheimer's disease based on adaptively extracted magnetoencephalogram rhythms. *Physiological Measurement*, **32**(8), 1163.

[35] Mapstone, M., Cheema, A. K., Fiandaca, M. S., Zhong, X., Mhyre, T. R., MacArthur, L. H., Hall, W. J., Fisher, S. G., Peterson, D. R., Haley, J. M., Nazar, M. D., Rich, S. A., Berlau, D. J., Peltz, C. B., Tan, M. T., Kawas, C. H. and Federoff, H. J. (2014). Plasma phospholipids identify antecedent memory impairment in older adults. *Nature Medicine*, **20**(4), 415–418.

[36] Jeong, J., Gore, J. C., and Peterson, B. S. (2001). Mutual information analysis of the EEG in patients with Alzheimer's disease. *Clinical Neurophysiology*, **112**(5), 827–835.

Chapter 9

SSA Application on Genetic Studies

9.1   Genomic Signal Processing

In 1953, James D. Watson and Francis Crick proposed the double helical structure of DNA (Deoxyribonucleic acid) through their landmark paper in the *Nature* journal [1]. Since then, analysis of DNA and extraction of a variety of information have become popular.

Since the genomic and proteomic data became available in the public domain, it has become increasingly important to be able to process this information for the benefit of clinicians and a wider public. In this context, traditional as well as modern signal processing methods have played an important role in these fields. This chapter aims at addressing these techniques and most importantly exploring the application of SSA to such data.

The genomic sequences may manifest themselves in a normal fashion where the protein coding regions (exons within genes) typically exhibit a periodic behaviour that is not found in other parts of the DNA molecule. On the other hand, the gene sequences are subject to change at some points due to crossover, and in the case of abnormalities, to mutation.

Traditional methods, though very popular, include application of frequency domain analysis to detect the cycle frequency of the DNA sequences by means of Fourier transform or other frequency estimation methods.

Some aspects of the DNA sequence analysis from the signal processing view point can be found in the introductory magazine article by Anastassiou [2]. Figure 9.1 demonstrates a simple schematic for part of a DNA molecule [3] with the double helix straightened out for simplicity. The four bases or nucleotides attached to the sugar phosphate backbone are denoted with the usual letters A, C, G, and T which stand respectively for adenine, cytosine, guanine, and thymine. Note that the base A always pairs with T, and C pairs with G. The two strands of the DNA molecule are therefore complementary to each other. The forward genome sequence corresponds to the upper strand of the DNA molecule. In this example this is ATTCATAGT. Note that the ordering is from the so-called 5' to the 3' end (left to right). The complementary sequence corresponds to the bottom strand, again read from 5' to 3' (right to left). This is ACTATGAAT in the example here. DNA sequences are always listed from the 5' to the 3' end since they are scanned in that direction when triplets of bases (codons) are used to signal the generation of amino acids.

Typically, in any given region of the DNA sequence, at most one of the two strands is active in protein synthesis (multiple coding areas, where both strands are separately active, are rare).

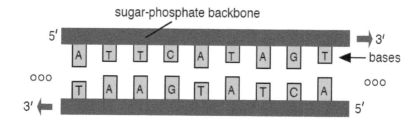

FIGURE 9.1: The DNA double helix (linearized schematic) [4].

Figure 9.2 shows the regions of interest in a DNA sequence. The genes are responsible for protein synthesis. Although all the cells in an organism have identical genes, only a selected subset is active in any particular family of cells. A gene, which for our purposes is a sequence made up from the four bases, is often divided into two sub-regions of exons and introns. (Procaryotes, which are cells without a nucleus, do not have introns). Only the exons are involved in protein-coding.

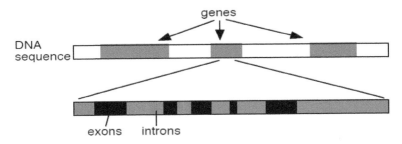

FIGURE 9.2: The genes and intergenic regions of a DNA sequence. The genes of eucaryotes have exons (protein coding regions) and introns [4].

In genomic sequences the bases in the exon region can be imagined to be divided into groups of three adjacent bases. Each triplet is called a *codon*. It is evident that there are 64 possible codons. Looking at the gene sequence from left to right, a codon sequence may be defined by concatenation of the codons in all the exons. Each codon (except the so-called stop codon) instructs the cell machinery to synthesise an amino acid. The codon sequence therefore uniquely identifies an amino acid sequence which defines a protein. Since there are 64 possible codons but only 20 amino acids, the mapping from codons to amino acids is many-to-one.

On the other hand, the introns do not participate in the protein synthesis because they are removed during the process by which the RNA molecules (called messenger RNA or mRNA) are formed. Thus, unlike the parent gene, the mRNA has no introns; it is a concatenation of the exons in the gene. The mRNA carries the genetic code to the protein machinery in the cell called the ribosome (located outside the nucleus). The ribosome produces the protein coded by the gene [4].

Genomics is an interesting field of study not only in biology but also in bioinformatics and signal processing.

In bioinformatics, sequence alignment is probably the most important topic. This is used to decide if two genes or proteins are related by function, structure, or evolutionary history, and can identify patterns of conservation and variability.

On the other hand, microarray technology provides a medium for matching known and unknown DNA samples based on hybridization (base-pairing). There are two major applications for this technology: (a) identification of a sequence (gene or gene mutation), and (b) determination of expression level (abundance) of genes.

Microarrays measure expression levels of thousands of gene simultaneously. Clustering of expression level data is one of the topics in machine learning which involves traditional statistical methods but also a graph-theoretic approach, an information-theoretic approach, and many other new methods. An example of hierarchical clustering methods can be seen in Figure 9.3 [5]. In this

Figure the rows show various gene expression levels and the columns represent the time progression.

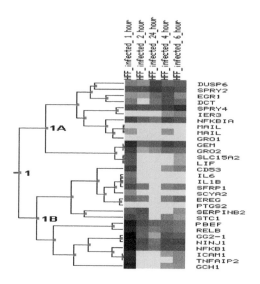

FIGURE 9.3: Hierarchical clustering of microarray expression using Random Forest [5].

Another important issue is the fusion of genomic data to enable classification of a protein, assuming that the following are known:

- original gene sequence encoding the protein,

- gene expression levels,

- some of the protein-protein interactions.

The classification problem is then how to combine various types of data to classify the protein.

9.1.1 Problems in Genomic Signal Processing

To apply DSP technique to genomic data, the DNA should be represented by numerical sequences. Usually, there are two options in doing that:

1. Create four binary sequences, one for each character (base), which specify whether a character is present (1) or absent (0) at a specific location. These are known as indicator sequences [6].

2. Assign meaningful real or complex numbers to the four characters A, T, G, and C. In this way, a single numerical sequence representing the entire character string is obtained.

Based on (a) any three of the four indicator sequences completely characterise the full DNA character string, and (b) the indicator sequences can be analysed to identify patterns in the structure of a DNA string.

The major signal processing domain activities in the context of genomic signal processing are therefore (*i*) detection of the periodic structure of the DNA sequences by finding out the fundamental frequencies of occurring A, C, T, and G, (*ii*) prediction of the structure of the DNA to model various normal or abnormal cases for each particular organism, (*iii*) clustering of gene expressions, (*iv*) classification of genes to discover the types of the organism, (*v*) detection of abnormalities in the form of anomalies in the DNA sequences. Consequently, there are often some optimisation problems to solve in the case of data fusion and multiple constraint systems.

9.1.2 Frequency Domain Analysis

Protein-coding regions of DNA have been observed to have a peak at frequency $2\pi/3$ in their Fourier spectra. This is called the period-3 property. This observation can be traced back to the 1980 work of Trifonov and Sussman [7]. The period-3 property is related to the different statistical distributions of codons between protein-coding and noncoding DNA sections and can be used as a basis for identifying the coding and non-coding regions in a DNA sequence.

For eucaryotes (cells with nucleus) this periodicity has mostly been observed within the exons (coding sub-regions inside the genes) and not within the introns (noncoding sub-regions in the genes). There are theories explaining the reason for such periodicity, but there are also exceptions to the phenomenon. For example, certain rare genes in S cerevisiae (also called baker's yeast) do not exhibit this periodicity [8]. Furthermore for procaryotes (cells without a nucleus), and some viral and mitochondrial base sequences, such periodicity has even been observed in noncoding regions [9].

For this and many other reasons, gene prediction is a very complicated problem (see the review article by Fickett [10]). Nevertheless, many researchers have regarded the period-3 property to be a good (preliminary) indicator of gene location. Techniques which exploit this property for gene prediction proceed by computing the discrete Fourier transform (DFT), which is expected to exhibit a peak at the (angular) frequency $2\pi/3$ due to the periodicity (e.g., see Figure 9.4 later). In fact this technique has also been used to isolate exons within the genes of eucaryotic cells [3, 8]. The periodic behaviour indicates strong short-term correlation in the coding regions, in addition to the long-range correlation or $1/f$-like behaviour exhibited by the DNA sequences in general. Although these methods seem to be robust, they have less flexibility and tenability than the adaptive systems.

The application of Fourier transform to DNAs and protein sequences results, however, was important in confirmation of the period-3 property in the protein-coding region. It is convenient to introduce indicator sequences for bases in DNA. For example the indicator for base A is a binary sequence of the form

$$x_A = 100110101000101010... \tag{9.1}$$

where 1 indicates the presence of an A and 0 indicates its absence and n represents the time sample. The indicator sequences for the other three bases are defined similarly. The same indicator can be found for G, T, and C respectively as $x_G(n), x_T(n)$ and $x_C(n)$. Consequently, their Fourier transforms within a window size N can be calculated separately as $X_A(k), X_G(k), X_T(k)$ and $X_C(k)$ where k represents the discrete frequency sample.

It has been noticed that protein-coding regions (exons) in genes have a period-3 component because of coding biases in the translation of codons into amino acids. This observation can be traced back to the research work of Trifonov and Sussman in 1980 [7]. The period-3 property is not present outside exons, thus can be exploited to locate exons. The overall power spectrum density (PSD) can be also found as:

$$S(k) = |X_A(k)|^2 + |X_G(k)|^2 + |X_T(k)|^2 + |X_C(k)|^2. \tag{9.2}$$

This is expected to have a peak at the sample value $k = N/3$ (corresponding to $2\pi/3$) for protein-coded regions. For long sequences a time-frequency approach using a short-term Fourier transform over a shorter sliding window or wavelet transform is preferable. This illustrates that $S(N/3)$ evolves along the DNA sequence. The window length N should be large enough so the periodicity effect dominates the background $1/f$ spectrum [7]. However, a long window implies larger computation complexity in predicting the exon location. The traditional filtering approaches can be followed to smooth $S(k)$ and provide a better spectrum of the peaks. Figure 9.4 shows an example.

Many other methods have been applied to genomic sequences to solve various aforementioned problems. One of the most interesting one is probably the work of Chang et al. [11] in the application of quaternions to jointly analyse the four sequences. Quaternion numbers [12] (also called the *hyper-complex numbers*) are the generalisation of complex numbers. In this method, it has been shown that the quaternion cross-correlation operation can be used to obtain both the global and local matching/mismatching information between two DNA sequences from the depicted one-dimensional curve and two-dimensional pattern, respectively.

9.2 Application

There are many different methods and techniques for analysing genetics data. Historically, such data have been analysed using parametric methods.

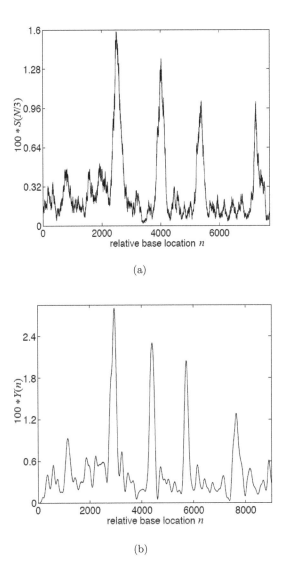

(a)

(b)

FIGURE 9.4: (a) The DFT based spectrum $S(N/3)$ for gene F56F11.4 in the C-elegans chromosome III and (b) the multistage narrowband bandpass filtered output for the same gene [4].

However, constraining pre-assumptions needed for parametric approaches led towards the growing popularity of nonparametric methods. Recently it has been concluded that non-parametric techniques can be used as an alternative approach for analysing genetics data because of their inherent nature [13], and accordingly the applications in biomedical and genetic fields have expanded. Among many non-parametric methods, SSA has proven to be a very successful tool in the analysis of genetic data and has illustrated its strong potential for such studies [14, 15].

The main advantages of the SSA technique in the field of genetics can be attributed to its signal extraction and filtering capabilities [16], batch processing of a set of a similar series [17] and derivation of an analytical formula of the signal [18]. The application of SSA for noise reduction in microarrays, and signal extraction in gene expression data have received much interest. The reason underlying the significant interest in the SSA technique's filtering capabilities are due to the fact that genetic data is often characterised by the existence of considerable noise; filtering this noisy data is considered to be one of the most arduous tasks when analysing genetic data [19, 20].

For example, microarray is a very useful method for acquiring quantitative data in genetics and researchers today are conducting most of their studies using this method. The main advantage of microarray is the capability of studying thousands of genes simultaneously. However, microarray data usually contains a high level of noise, which can reduce the performance of the algorithms [21].

More interestingly, a general similarity between Colonial theory (CT) and SSA and how the different steps of that theory are fully consistent with the procedure underlying SSA has been found which is described in detail in the following sections of this chapter.

Later on in this chapter, the main similarities between CT and SSA, the application of SSA for signal extraction, theoretical developments leading to what is termed 'two-dimensional SSA' , the application of SSA based on minimum variance and a hybrid modelling approach combining SSA and an autoregressive (AR) model which have enabled researchers to achieve enhanced results alongside an improved estimation of extracted signals are discussed in depth.

9.3 Singular Spectrum Analysis and Colonial Theory

Presented below is an outline of CT along with the steps underlying the SSA process in order to show the similarities in Figure 9.5. As visible through this figure, both CT and SSA have their origins or inputs in univariate form of a single cell in the case of CT and a one dimensional time series in the case of SSA. Thereafter at the second stage, via a procedure known as embedding

SSA develops a multi-dimensional time series which corresponds to the colony of cells which appears in the CT. In the third stage of the CT the colony of cells develops further into specialized cells aiming at performing certain tasks, and at the same stage in SSA the SVD procedure gives us the eigenvalues which provide crucial information relating to the time series enabling differentiation between signal, harmonic components and noise which can be viewed as the specialized tasks performed by the specialised cells in CT. In the final stage, the specialised cells in CT form a single multicellular organism whilst the selected components in SSA which correspond to signal and harmonics provide the filtered less noisy time series.

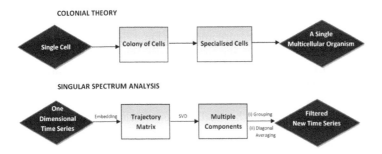

FIGURE 9.5: The CT and SSA models which clearly reflect their similarities.

It is widely accepted that the first grouping in nature happened when multicellular organisms arose from a single cell and generated a multi-celled organism [22]. In the beginning of life there were only single cells. Today, after many millions of years, most animals, plants, fungi, and algae are made up of multiple cells that work together as a single being [23]. There are various mechanisms that may be able to explain how multicellularity could have evolved, and among them, CT is most credited by developmental scientists [24]. CT was introduced by Haeckel in 1874 and later modified by Metschnikoff (1886) and revived by Hyman (1940) [25]. As explained in [26], this theory claims that through the association of many one-celled flagellates individuals form a colony. Thereafter, by increasing the number of individual cells they become more specialised in both structure and functions, and this process continues till the individuality in the cells is lost and the whole colony becomes a single multicellular individual. In what follows, the brief explanation given above through Figure 9.5 has been expanded by describing in more detail how the different steps of CT are fully consistent with the procedure underlying SSA. As the objective is to provide a clear picture on the similarities between SSA and CT, the detailed explanation under the stages and steps of the SSA techniques is presented below.

9.3.1   Stage 1: Decomposition

In SSA, the procedure begins with a one dimensional time series, $Y_N = (y_1, \ldots, y_N)$ where N is the length of the series. Here Y_N can be considered as an individual or single cell which is the starting point of CT. Thereafter, over time this single cell evolves and generates a multi-celled organism. The SSA technique consists of two choices, the window length L and the number of eigenvalues r [27]. At the first stage, the single SSA choice that needs to be determined is L. In SSA the number of components are related to the selection of the proper window length L which should be defined such that it minimises the signal distortion and maximises the residual noise level. Likewise, in nature there is a limit on the number of cell types in an organism developed from a single cell. It is assumed that the number of cell types is determined by the balance between selective pressure and functional requirements, whilst variety is favoured by selection, functional needs limit the number of cell types [28]. However, in SSA a general rule in selecting L for different time series with different structure has not been imposed. For example, in places where there is a periodic component with an integer period like a seasonal component, to obtain a higher separability the tradition is to select L proportional to that period [29]. Likewise in CT, different kinds of organisms necessitate different numbers of cell types and for a specific size, the number of cell types in animals is greater than for green plants, and this difference is consistent with the relation between complexity and motility [30].

1st Step: Embedding

Embedding is one of the four steps in SSA in which mapping a one dimensional time series Y_N creates a multi-dimensional variable of $X_1, \ldots, X_K$. It is clear that this can be viewed as the creation of the colony from the initial single cell because by taking the Y_N, multiple dimensions from the same series have been created.

It should be noted that unlike the Symbiotic theory [31], which assumes that the symbiosis of various species caused a multicellular organism, in CT it is the symbiosis of many cells of the same species that forms a multicellular organism. This point is interesting, as it can be referred to as the first and main difference between SSA and PCA. In the latter, the obtained matrix is achieved by considering different time series (multiple cells) whilst in SSA one time series has been considered (single cell).

Moreover, transferring a one dimensional time series into a trajectory matrix will enable us to significantly reduce the computation time required for running the algorithm. Furthermore, by analysing the eigenvalues with the aim of filtering the signal and noise, the signal to noise ratio will be optimised in the reconstructed new time series. Likewise, increasing in size is initially favoured by individual cells since multicellular organisms do not have the size limit which is mainly imposed by diffusion. As the surface-to-volume ratio decreases in a given single cell, with increased size they will have difficulty in getting the required nutrients and transporting the cellular waste products out the cell [26, 28].

2nd Step: Singular Value Decomposition

As mentioned in Chapter 2, singular value decomposition is the second of the four steps in SSA. At the outset it is noteworthy that according to CT, as what happens in the blastula of modern animals, by increasing the interdependency level in a colony some of the cells specialise to do different tasks. Consequently, by increasing the number of specialised cells, early animals developed more complex bodies just like the gastrulae of modern animals. Finally by obtaining an ever more complex level, cells formed tissues and then organs [28].

SVD is a procedure which is performed on $\mathbf{X}$ and can be seen as a matrix that contains the entire colony of cells generated from the original single cells. The SVD step provides us with eigenvalues relating to the single cell where those cells that have now begun specialising emerge. Given that each of these cells are to aid in performing a certain task, a direct link can be created between these structurally and functionally differentiated cells and different components obtained via SVD and identify them as either trend, periodic, quasi-periodic component, or noise. This leads us to the second stage of SSA and its final two steps.

9.3.2 Stage 2: Reconstruction

In SSA, this is the stage where the eigenvalues extracted via the SVD step are analysed in order to differentiate between noise and signal in a time series. In CT this would correspond to identifying which of the specialized cells are able to successfully carry out the reproductive task and which cells are responsible for viability. As soon as the leading signal components have been identified in a time series, the series components can be recovered through the reconstruction stage. However, there are always some components with less frequency that cannot be considered as the dominant component for reconstruction, but may still help us in obtaining a better estimation of the series. These components normally appear together with the leading components in the same group. Periodogram analysis of the original series and eigenvectors can aid us immensely in achieving the most accurate grouping here [29].

1st Step: Grouping

It is widely accepted that the first grouping in nature happened when multicellular organisms arose from a single cell and generated a multi-celled organism [22]. At the very beginning of life there were only single cells. Today, after millions of years, most animals, plants, fungi, and algae are made up of multiple cells that work together as a single being [23].

Grouping is the third out of the four steps in SSA. This is one of the most important steps as the quality of the filtering achieved via SSA depends on the successful analysis of eigenvalues and selection of appropriate groups of them to rebuild the less noisy time series. In brief, this step involves grouping together the eigenvalues with similar characteristics, i.e. signal and harmonic components, whilst leaving out those corresponding to noise. The similarity with CT lies in the fact that out of the specialised cells, certain cells are successful in carrying out their reproduction task whilst certain other cells fail along the way. Accordingly, grouping can be helpful in identifying the productive and counterproductive cells so that a more accurately prediction of which particular cells form organs that are vital for performing different tasks in a whole body can be made.

2nd Step: Diagonal Averaging

The diagonal averaging step in SSA transforms the matrix of grouped eigenvalues back into a Hankel matrix which can later be converted into a time series. The resulting time series will be the less noisy, filtered time series corresponding to the original one-dimensional time series that was applied to the SSA process at the beginning. This step is important and similar to the final stage of CT where a single multiple organism is formed. This is because in the grouping step the productive cells should be selected out of the whole cells to compensate for the increasing cost of reproduction to survival with increasing colony size.

It should be noted that there are opportunities and costs associated with the larger size obtained by unicellular-multicellular transition. One of the main advantages is the ability of sharing tasks. Multicellular organisms are able to share complementary tasks between different cells simultaneously [32, 33]. For example, during the photosynthesis in the presence of oxygen, nitrogen fixation does not adequately catalyse the fixation process, hence it should happen either in a different time or place [34].

Likewise, in analysing a time series by raising the number of observations, different types of components describing the series might be found, for example, if by increasing the length of the series a better estimation of the series by observing harmonic and cyclic components in addition to trend and noise components may be obtained.

After achieving the major functional specialisations, there would presumably be a decline in adding capabilities obtained by additional cell types [28]. Similarly, in a time series, after extracting different components related to the trend, oscillation and noise, increasing the number of observations provides more components but all of these will be categorised in a different group of components as mentioned before. This matter in CT can be considered as the role of a small variant. According to this role, the number of cell types grows with two factors: body length and log of number of cells [35]. However regarding the latter case, the slope is only 0.056 which explains that adding a single cell will have a slight impact on the performance of a large organism [36].

9.4 Signal Extraction and Filtering

This section presents the applications of SSA for signal extraction and noise filtering in genetics. The first such application was reported in 2006 where SSA was used for signal extraction of *Drosophila melanogaster*'s gene expression profile [16]. The idea of using SSA for signal extraction was then followed in an approximately similar study in [37] which led to an improved result. By 2008 a more technical study conducted on the methodology of signal extraction from the noisy *Bicoid* (*Bcd*) [1] protein profile in *Drosophila melanogaster* was presented in [18]. The problem under investigation in that study was complicated by two facts: (*i*) the data contained outliers and (*ii*) the data was exceedingly noisy and the noise consisted of an unknown structure. The author examined two approaches for reconstructing the signal more precisely: the use of small window length and the improvements to separability by adding a constant to the series.

In addition, the activation of the *hunchback* (*hb*) gene in response to different concentrations of *Bcd* gradient was studied in [38] and SSA was applied for filtering two kinds of noise; experimental noise and the noise caused by variability in the nuclear order [38].

9.5 SSA Based on Minimum Variance

Recently, a modified version of SSA was examined for filtering and extracting the *bcd* gene expression signal [39] and the results illustrated that SSA based on minimum variance, SSA_{MV}, can significantly outperform the previous methods used for filtering noisy *Bcd* [39].

SSA based on minimum variance mainly relies on the concept that by dividing the given noisy time series into the mutually orthogonal noise and signal + noise components, an enhanced estimation of the signal can be achieved. Thus, after performing SVD, by adapting the weights for different obtained

---

[1]The italic lower-case *bcd* represents either the gene or mRNA and *Bcd* refers to the protein. This can be applied for all the other genes mentioned in this chapter (for example, *hb* and *Hb*).

singular components, an estimation of the Hankel matrix $\mathbf{X}$, will be achieved which in turn is corresponds to a filtered series.

Figure 9.6 shows the signal extracted by SSA_{MV} [39] along with a fitted trend obtained by the SDD model as the most widely accepted model in analysing Bcd gradient. A detailed description of the SSA based on minimum variance was given in Chapter 2 (for more details, see also [39, 40]).

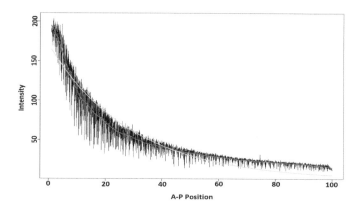

FIGURE 9.6 (See color insert.): Temporal dynamics of Bcd expression. Red and green lines show the trend respectively extracted by SSA_{MV} and SDD [39].

9.6   SSA Combined with AR Model

Study of rhythmic cellular processes is an important step towards the understanding and applications of various microarray processing techniques. Rhythmic cellular processes are mainly regulated by different gene products, and can be measured by using multiple DNA microarray experiments. The characteristics of the rhythmic gene expression is as follows [41]: the number of time points and cycles related to a profile is usually very few. For example, one cell-cycle may be sampled/observed in only 14 time instants [17]. This dataset usually contains many missing values which need to be determined [42], the intervals spaced between time points are not equal and the gene expression data is extremely noisy [41].

The SSA technique has been employed for extracting the dominant trend from the noisy expression profiles after reducing the effect of noise level [14]. A combination of the autoregressive (AR) model and SSA has been utilised for analysing the periodicity of the transcriptome of the intraerythrocytic developmental cycle. The combination enabled to successfully identify up to

90% of genes (4496 periodic profiles) in *P. falciparum*, which is a considerable achievement in terms of detecting 777 additional periodic oligonucleotides in comparison to the results provided in [43]. Figure 9.7 depicts the improvement yielded by using SSA in analysing periodicity.

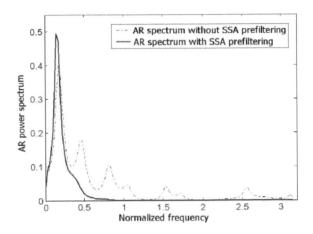

FIGURE 9.7: The AR spectra of the expression profile of Dihydrofolate Reductase-Thymidylate Synthase with and without SSA filtering [14].

Four subsequent research studies followed a similar procedure and successfully improved the capability of detecting periodicity from 60% to 80% [44]–[47]. According to [41], the periodic profiles can be detected by combining the SSA and AR models as follows:

- SSA is initially employed for the filtering of each expression data. To this end, only those expression profiles where the sum of first two eigenvalues over the sum of all eigenvalues is greater than 0.6 are selected for reconstruction, filtering, and further processing.

- The second step is devoted to estimation of the AR spectrum, frequency f_i at a peak value point, and the ratio of the power f_i regions of interest (ROI) of the reconstructed profiles selected in the first step. The ratio between the signal power within the frequency band [f_{i-1}, f_{i+1}], and the total power is estimated using the so-called power ratio $PR = \frac{power_i}{power_{total}}$, where $power_i$ is related to frequency f_i.

- If the estimated power ratio PR is larger than 0.7, the corresponding profile is classified as periodic [41].

Let us now briefly discuss how an autoregressive model of order p $(AR_{(p)})$ can be used after the SSA analysis is performed on the original noisy series. Let the microarray data be $g \times N$ matrix $(g \gg N)$, where g is the number of gene expression profiles and N corresponds to the number of samples. Time series gene expression $Y_N = (y_1, \ldots, y_N)$ can then be written in the form of an $AR_{(p)}$ model by forward-backward linear prediction as follows [47]:

$$y_i = \alpha_1 y_{i-1} + \alpha_2 y_{i-2} + \ldots + \alpha_p y_{i-p} \qquad (i = p+1, \ldots, N). \qquad (9.3)$$

Using the forward prediction linear system the AR coefficients $(\alpha_1, \alpha_2, \ldots, \alpha_p)$ can be estimated and the gene expression can be recognised as a linear system. In an $AR_{(p)}$ model, p has to be set in a way that the system models the desired trend and avoids redundant data such as noise s performed using the Akaike method previously. By choosing m gene expression profiles, the number of linear equations is equal to $m \times 2(N - p)$.

As mentioned above, the SSA technique is often applied prior to spectral analysis, as a filtering method, in order to achieve better accuracy. This is done by ignoring the small singular values in the reconstruction stage. Removing the noise component from the original noisy signal Y_N yields the noise reduced series $\hat{Y}_N$ which can consequently be used for estimating the AR coefficients α_i $(i = 1, \ldots, p)$. The value of α_i is generally estimated using Yule–Walker equations [47].

9.7   Application of Two Dimensional SSA

Bcd is a transcriptional regulator of downstream segmentation genes where the alteration in *Bcd* gradient shifts the downstream patterns [51]. However, it has been accepted that in the embryos, zygotic gene products are considerably more precisely positioned than the products related to maternal genes, which indicates an embryonic error depletion process [52]. In 2011 the anterior-posterior (A-P) segmentation of *Drosophila* was studied in [53] to determine how gene regulation dynamics control noise. In this research the activation of the *hb* gene by the *Bcd* protein gradient in the anterior part was studied by modelling the noise observed in *hb* regulation using a chemical master equation approach. For solving this model, the MesoRD software package has been used which mainly follows a stochastic approach [54], and the results indicate that *Hb* output noise is mostly dependent on the transcription and translation dynamics of its own expression, and that the multiple *Bcd* binding sites located in the *hb* promoter also improve pattern formation [55].

The 2D SSA can also be used to measure the between-nucleus variability (noise) seen in the gradient of *Bcd* in *Drosophila* embryos. The 2D SSA is utilized for determining the noise for comparing the results of fixed immunostained embryos with live embryos with fluorescent *bcd* (bcd-GFP) [56]. As can be seen in Figure 9.8 the results indicate that the nucleus-to-nucleus noise in *Bcd* intensity, is signal-independent for live and fixed immunstained embryos more sensitive to the nuclear masking technique which is used to extract the intensity values. An advantage of using 2D SSA over classical methods or such application is that the immunochemical intensification of the signal in fixed embryos via the secondary antibodies would not affect the noise statistics considerably [56].

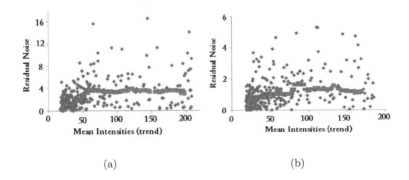

(a)   (b)

FIGURE 9.8: Between-nucleus noise and its corresponding signal dependency at different ROI, between-nucleus noise is mostly correlated with the signal level. a) Nuclear intensity of *Bcd*, obtained by small ROI, and b) *Bcd* nuclear intensity with large ROI [56].

9.8   Eigenvalues Identification in SSA Signal Extraction

As stated earlier, even though the signal extraction by SSA appears to be simple, it might be a complicated task in practice, particularly for *bcd*, since the signal cannot be separated from noise by selecting a few leading eigenvalues. This issue, has been addressed in a few studies. For instance, the use of either a small window length or adding a constant to the series for improving the separability between signal and noise is recommended in [18]. However, such an approach may not be suitable. Accordingly, there is a need for a general approach that is not sensitive to series structure, noise level and sample size to some extent.

A more general approach to address this issue was studied in [57, 58]. The method has been used mainly for noise reduction, filtering, signal extraction and distinguishing chaos from noise in time series analysis [59]. The number of the eigenvalues is identified based on their distribution [57] rather than checking the properties of each eigenvalue or constructed component by the corresponding eigenvalue separately. A brief description of the algorithm along with its application for signal extraction of four different genes; *bcd*, *caudal* (*cad*), *giant* (*gt*) and *even-skipped* (*eve*), which are among the most important zygotic segmentation genes, are studied. The approach enables one to decide the appropriate number of eigenvalues related to the gene signal.

The algorithm is composed of two complementary stages. In the first stage, the optimal value of r is determined for splitting signal and noise, and the noise reduced signal is reconstructed in the second stage similar to the second stage of SSA.

1. Compute the matrix $\mathbf{A} = \mathbf{X}\mathbf{X}^T/tr(\mathbf{X}\mathbf{X}^T)$.

2. Compute the eigenvalues and eigenvectors of the matrix $\mathbf{A}$ and represent it in the form $\mathbf{A} = \mathbf{P}\mathbf{\Gamma}\mathbf{P}^T$. Here, $\mathbf{\Gamma} = diag(\zeta_1, \ldots, \zeta_L)$ is the diagonal matrix of the eigenvalues of $\mathbf{A}$ that has the order $(1 \geq \zeta_1 \geq \zeta_2, \ldots, \zeta_L \geq 0)$ and $\mathbf{P} = (P_1, P_2, \ldots, P_L)$ is the corresponding orthogonal matrix of the eigenvectors of $\mathbf{A}$.

3. Simulate the original series m times and calculate the eigenvalues for each m series. The simulation of y_i is generated from a uniform distribution with boundaries $y_i - a$ and $y_i + b$, where $a = \mid y_{i-1} - y_i \mid$ and $b = \mid y_i - y_{i+1} \mid$.

4. Calculate the coefficient of skewness, $skew(\zeta_i)$, and the coefficient of variation, $CV(\zeta_i)$, for each eigenvalue. If $skew(\zeta_c)$ and $CV(\zeta_i)$ are the maximum, then select $r = c - 1$. This can split the eigenvalues into two groups, from ζ_1 to ζ_{c-1} which correspond to the signal, and the rest to the noise component. The noise part has almost a U shape.

It is recommended to calculate the absolute values of the correlation between ζ_i and ζ_{i+1}, and plot them in one figure. Similarly, if $\rho(\zeta_{c-1}, \zeta_c)$ is the minimum, and the pattern for $\rho(\zeta_c, \zeta_{c+1})$ to $\rho(\zeta_{L-1}, \zeta_L)$ has the same pattern for the white noise, then choose $r = c - 1$.

Let us now consider the application of the above algorithm to four different segmentation genes namely *bcd, cad, gt* and *eve* which among them *bcd* and *cad* are maternal and *gt* and *eve* are respectively related to the gap and pair rule categories of zygotic genes.

The *bcd* expression is completely maternal which its distribution begins at about cleavage cycle 9. Figure 9.9(a) depicts a typical example of *Bcd* gradient related to cleavage cycle 14(3).

The *cad* mRNA has both maternal and zygotic origin and the maternal transcripts begin to distribute immediately after fertilization. However, proteins encoded by *gt* and *eve* were reported to appear at cycle 12 and 10 respectively and it is accepted that the posterior stripe of *gt* expression is regulated by *bcd* and *cad*.

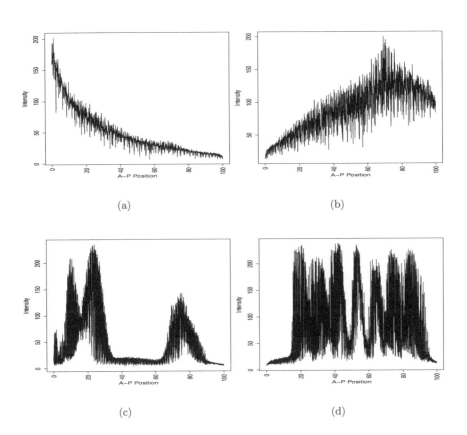

FIGURE 9.9: Experimental data from *Drosophila melanogaster* embryo; (a): *bcd*, (b): *cad*, (c): *gt*, and (d): *eve*.

To apply the above algorithm, several copies of each gene need to be generated. Then, different embryos for each gene are selected. Accordingly, each copy of gene expression data is simulated m times (here $m = 10^4$ is considered) and the distribution ζ_i are obtained. Studying the distribution of ζ_i provides the necessary guidelines on how to select a proper value for r in reconstruction stage. In the following, the data for each gene is analysed and discussed in more detail whilst the results of the other data sets are summarised based on the outcomes of the skewness, variation and correlation coefficients. The window length $L = 200$ has been used for analysing the *bcd*, *cad*, *gt* and *eve* genes.

Figure 9.10 illustrates the results of $skew(\zeta_i)$ (left) and $CV(\zeta_i)$ (right). As it appears from Figure 9.10, the maximum value of $skew$ is attained for ζ_2 for both *bcd* and *cad*, whereas, $skew(\zeta_4)$ is maximum for *eve* and *gt*. The results of CV splits the eigenvalues into two groups; the second group looks like a U shape which is related to the noise component. The results indicate that $r = 2, 2, 3, 3$ are acceptable choices for extracting the *bcd*, *cad*, *eve* and *gt* signal, respectively. Furthermore, the result of the Spearman correlation, ρ, can be used as an auxiliary information if the $skew$ and CV provide different results. However, in these typical examples, the results are the same which are also supported by the results of the correlation coefficient. The results confirm that the minimum value of ρ are emerged between (ζ_2, ζ_3), (ζ_2, ζ_3), (ζ_3, ζ_4) and (ζ_3, ζ_4) for *bcd*, *cad*, *bcd* and *cad*, respectively. This shows $r = 2, 2, 3, 3$ are reasonable choices (see Figure 9.11).

In the next step, after identifying r, the selected components are used in the second stage of SSA (grouping and diagonal averaging) to reconstruct the noise-reduced gene data. Figure 9.12 shows the result of the gene signal extraction which is a less noisy signal than the original one. The red and the black lines correspond to the reconstructed signal and the original signal, respectively. The estimated less noisy signal confirms that the r selected by the above algorithm yields a proper signal extraction results for all cases.

9.9 Concluding Remarks

Even though the signal extraction of gene expression profiles appears to be simple, in practice it is an arduous and complicated task considering the existence of non stationary noise. The feasibility of capturing the signal, and filtering the noise of gene expression data related to some model organisms presented in this chapter suggests that SSA may be of general use in evaluating other expressional systems.

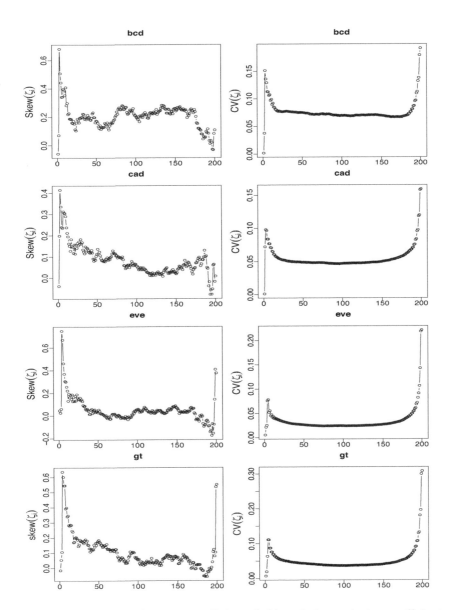

FIGURE 9.10: The skewness coefficient (left) and the variation coefficient of ζ_i (right) for the first series of *bcd, cad, gt* and *eve* data.

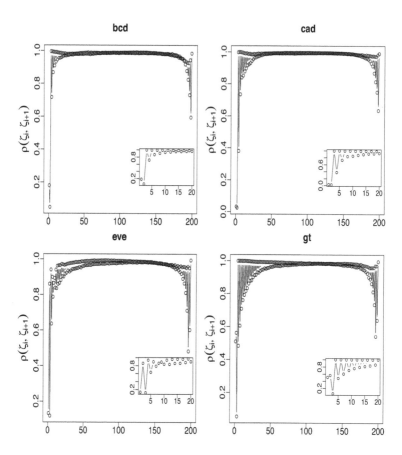

FIGURE 9.11: The correlation between ζ_i and ζ_{i+1} for the first series from each data.

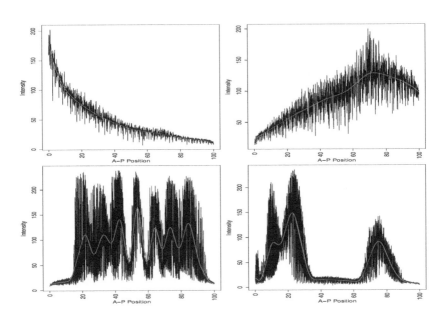

FIGURE 9.12 (See color insert.): Original (black) and reconstructed (red) series.

Bibliography

[1] Watson, J. D., and Crick, F. H. (1953). Molecular structure of nucleic acids. *Nature*, **171**(4356), 737–738.

[2] Anastassiou, D. (2001). Genomic signal processing. *Signal Processing Magazine*, **18**(4), 8–20.

[3] Alberts, B., Bray, D., Hopkin, K., Johnson, A., Lewis, J., Raff, M., Robert, K., and Walter, P. (2013). *Essential Cell Biology*. Garland Science.

[4] Vaidyanathan, P. P., and Yoon, B. J. (2004). The role of signal-processing concepts in genomics and proteomics. *Journal of the Franklin Institute*, **341**(1), 111–135.

[5] http://www.genomatix.de

[6] Tiwari, S., Ramachandran, S., Bhattacharya, A., Bhattacharya, S., and Ramaswamy, R. (1997). Prediction of probable genes by Fourier analysis of genomic sequences. *Bioinformatics*, **13**(3), 263–270.

[7] Trifonov, E. N., and Sussman, J. L. (1980). The pitch of chromatin DNA is reflected in its nucleotide sequence. *Proceedings of the National Academy of Sciences*, **77**(7), 3816–3820.

[8] Brown, P. O., and Botstein, D. (1999). Exploring the new world of the genome with DNA microarrays. *Nature Genetics*, **21**, 33–37.

[9] Cosic, I. (1994). Macromolecular bioactivity: is it resonant interaction between macromolecules?–theory and applications. *Biomedical Engineering, IEEE Transactions*, **41**(12), 1101–1114.

[10] Fickett, J. W. (1996). The gene prediction problem: an overview for developers. *Computers Chemistry*, **20**(1), 103–118.

[11] Chang, H. T., Kuo, C. J., Lo, N. W., and Lv, W. Z. (2012). DNA sequence representation and comparison based on quaternion number system. *International Journal of Advanced Computer Science and Applications (IJACSA)*, **3**(11).

[12] Kantor, I. L., and Solodovnikov, A. S. (1989). *Hypercomplex numbers: an elementary introduction to algebras*. **302**. New York: Springer-Verlag.

[13] Golyandina, N., Nekrutkin, V., and Zhigljavsky, A. A. (2001). *Analysis of time series structure: SSA and related techniques*. CRC press.

[14] Du L. P., Wu S.H., Liew A. W. C., Smith D. K., and Yan H. (2008). Spectral Analysis of Microarray Gene Expression Time Series Data of Plasmodium Falciparum. *International Journal of Bioinformatics Research and Applications*, **4**(3), 337–349.

[15] Tang, V. T., and Yan, H. (2012). Noise reduction in microarray gene expression data based on spectral analysis. *International Journal of Machine Learning and Cybernetics*, **3**(1), 51–57.

[16] Holloway, D. M., Harrison, L. G., Kosman, D., Vanario-Alonso, C. E., and Spirov, A. V. (2006). Analysis of Pattern Precision Shows that Drosophila Segmentation Develops Substantial Independence from Gradients of Maternal Gene Products. *Developmental Dynamics*, **235**(11), 2949–2960.

[17] Spellman, P. T., Sherlock, G., Zhang, M. Q., Iyer, V. R., Anders, K., Eisen, M. B., Brown, P. O., Botstein, D., and Futcher, B. (1998). Comprehensive identification of cell cycle-regulated genes of the yeast Saccharomyces cerevisia by microarray hybridization. *Molecular Biology of the Cell*, **9**(12), 3273–3297.

[18] Alexandrov, T., Golyandina, N., and Sprov, A. (2008). *Singular Spectrum Analysis of Gene Expression Profiles of Early Drosophila embryo: Exponential-in-Distance Patterns*. Hindawi Publishing Corporation.

[19] Gregor, T., Wieschaus, E. F., McGregor, A. P., Bialek, W., and Tank, D. W. (2007). Stability and nuclear dynamics of the bicoid morphogen gradient. *Cell*, **130**(1), 141–152.

[20] Hilfinger, A., and Paulsson, J. (2011). Separating intrinsic from extrinsic fluctuations in dynamic biological systems. *Proceedings of the National Academy of Sciences*, **108**(29), 12167–12172.

[21] Klebanov, L., and Yakovlev, A. (2007). How high is the level of technical noise in microarray data. *Biol Direct*, **2**(9).

[22] Michod, R. E. (2007). Evolution of individuality during the transition from unicellular to multicellular life. *Proceedings of the National Academy of Sciences*, **104**(Suppl 1), 8613–8618.

[23] Adl, S.M., Simpson, A.G., Farmer, M. A., Andersen, R. A., Anderson, O. R., Barta, J. R., et al. (2005). The new higher level classification of eukaryotes with emphasis on the taxonomy of protists. *Journal of Eukaryotic Microbiology*, **52**(5), 399–451.

[24] Wolpert, L., Szathmáry, E. (2002). Multicellularity: evolution and the egg. *Nature*, **420**(6917), 745–745.

[25] Kotpal, R. L. (2012). *Modern Text Book of Zoology: Invertebrates*. Rastogi Publications.

[26] Kirk, D. L. (2005). A twelve step program for evolving multicellularity and a division of labor. *BioEssays*, **27**(3), 299–310.

[27] Hassani, H., and Mahmoudvand, R. (2013). Multivariate Singular Spectrum Analysis: A general view and new vector forecasting approach. *International Journal of Energy Statistics*, **1**(1), 55–83.

[28] Grosberg, R. K., and Strathmann, R. R. (2007). The evolution of multicellularity: a minor major transition? *Annual Review of Ecology, Evolution, and Systematics*, 621–654.

[29] Hassani, H. (2007). Singular Spectrum Analysis: Methodology and Comparison. *Journal of Data Science*, **5**, 239–257.

[30] Cavalier-Smith, T. (1991). Cell diversification in heterotrophic flagellates. *The biology of free-living heterotrophic flagellates*, **113**, 131.

[31] Roberts, L. S., Miller, F., and Hickman Jr, C. P. (2001). *Integrated Principles of Zoology*. McGraw Hill, Boston.

[32] Bonner, J. T. (2004). Perspective: the size-complexity rule. *Evolution*, **58**(9), 1883–1890.

[33] Queller, D. C. (2000). Relatedness and the fraternal major transitions. *Philosophical Transactions of the Royal Society of London. Series B: Biological Sciences*, **355**(1403), 1647–1655.

[34] Kaiser, D. (2001). Building a multicellular organism. *Annual review of genetics*, **35**(1), 103–123.

[35] Bonner, J. T. (1965). *Size and Cycle*. Princeton, NJ: Princeton Univ. Press.

[36] Bell, G., and Mooers, A. O. (1997). Size and complexity among multicellular organisms. *Biological Journal of the Linnean Society*, **60**(3), 345–363.

[37] Surkova, S., Kosman, D., et al. (2008). Characterization of the Drosophila segment determination morphome. *Developmental Biology*, **313**(2), 844–862.

[38] Lopes, F. J. P., Vieira, F. M. C., Holloway, D. M., Bisch P. M., and Spirov, A. V. (2008). Spatial Bistability Generates hunchback Expression Sharpness in the Drosophila Embryo. *PLoS Comput Biol*, **4**(9), e1000184.

[39] Hassani, H., and Ghodsi, Z. (2014). Pattern Recognition of Gene Expression with Singular Spectrum Analysis. *Medical Sciences*, **2**(3), 127–139.

[40] Hassani, H. (2010). Singular spectrum analysis based on the minimum variance estimator. *Nonlinear Analysis: Real World Applications*, **11**(3), 2065–2077.

[41] Liew A. W. C, and Yan, H. (2009). Reliable Detection of Short Periodic Gene Expression Time Series Profiles in DNA Microarray Data. *IEEE International Conference on Systems, Man, and Cybernetics*. San Antonio, TX, USA.

[42] Gan, X., Liew, A. W. C., and Yan, H. (2006). Microarray Missing Data Imputation based on a Set Theoretic Framework and Biological Consideration. *Nucleic Acids Research*, **34**(5), 1608–1619.

[43] Bozdech, Z., Llinás, M., Pulliam, B. L., Wong, E. D., Zhu, J., and DeRisi, J. L. (2003). The transcriptome of the intraerythrocytic developmental cycle of Plasmodium falciparum. *PLoS biology*, **1**(1), e5.

[44] Tang, T. Y., and Yan, H. (2010). Identifying periodicity of microarray gene expression profiles by autoregressive modelin and spectral estimation. *Ninth International Conference on Machine Learning and Cybernetics*, 3062–3066.

[45] Vivian, T. Y. T., Liew, A. W. C., and Yan, H. (2010). Periodicity analysis of DNA microarray gene expression time series profiles in mouse segmentation clock data. *Statistics and Its Interface*, **3**(3), 413–418.

[46] Tang, T. Y., Liew, A. W. C, and Yan, H. (2010). Analysis of Mouse Periodic Gene Expression Data Based on Singular Value Decomposition and Autoregressive Modelling. it International Association of Engineers (IAENG), **1**.

[47] Tang, V. T., and Yan, H. (2012). Noise reduction in microarray gene expression data based on spectral analysis. *International Journal of Machine Learning and Cybernetics*, **3**(1), 51–57.

[48] Schott, J. R. (2005). *Matrix analysis for statistics.* Wiley series in probability and statistics. Wiley-Interscience, Hoboken.

[49] Zhang, J., Hassani, H., Xie, H., and Zhang, X. (2014). Estimating multi-country prosperity index: A two-dimensional singular spectrum analysis approach. *Journal of Systems Science and Complexity*, **27**(1), 56–74.

[50] Golyandina, N. E., and Usevich, K. D. (2010). 2D-Extension of singular spectrum analysis: Algorithms and elements of theory. *Matrix Methods: Theory, Algorithms*, 449–473

[51] Porcher, A., and Dostatni, N. (2010). The bicoid morphogen system. *Current Biology*, **20**(5), R249–R254.

[52] Spirov, A. V., and Holloway, D. M. (2003). Making the body plan: precision in the genetic hierarchy of Drosophila embryo segmentation. *In silico biology*, **3**(1), 89–100.

[53] Holloway, D. M., Lopes, F. J., da Fontoura Costa, L., Traven?olo, B. A., Golyandina, N., Usevich, K., and Spirov, A. V. (2011). Gene expression noise in spatial patterning: hunchback promoter structure affects noise amplitude and distribution in Drosophila segmentation. *PLoS computational biology*, **7**(2), e1001069.

[54] http://mesord.sourceforge.net

[55] Holloway, D. M., and Spirov, A. V. (2011). Gene Expression Noise in Embryonic Spatial Patterning. *21st International Conference on Noise and Fluctuations*.

[56] Golyandinaa, N. E., Holloway, D. M, Lopesc, F. J. P. (2012). Measuring gene expression noise in early Drosophila embryos: nucleus-to-nucleus variability. *Procedia Computer Science*, **9**, 373–382.

[57] Hassani, H., Alharbi, N., and Ghodsi, M. (2014). A short Note on the Pattern of Singular Values of a Scaled Random Hankel Matrix. *International Journal of Applied Mathematics*, **27**(3), 237–243.

[58] Hassani, H., Alharbi, N., and Ghodsi, M. (2014). A study on the empirical distribution of the scaled Hankel matrix eigenvalues. *Journal of Advanced Research*, doi:10.1016/j.jare.2014.08.008.

[59] Hassani, H., Alharbi, N., and Ghodsi, M. (2014). Distinguishing chaos from noise: A new approach. *International Journal of Energy and Statistics*, **2**(2), 137–150.

[60] Golyandina, N., and Zhigljavsky, A. (2013). *Singular Spectrum Analysis for Time Series*. Springer Briefs in Statistics. Springer.

Chapter 10

Conclusions and Suggestions for Future Research

Singular spectrum analysis in its various forms (multivariate, two-dimensional, tensor-based, hypercomplex, etc.) is largely dependent on the fundamental concept of subspace decomposition which has roots in the early mathematical development of signal, image, and information processing. In an intuitive and simple way, it introduces and involves the so-called embedding dimension in order to give freedom to the researcher to expand the number of subspaces to at least the number of actual components of interest. This is probably the main and most favourable concept which makes it different from its parent, principal component analysis.

As we moved on in preparation of the materials, we expanded this simple concept into cases where we can have trends correlated or dependent on each other, and so we further benefit from merging their Hankel matrices. On the other hand the two dimensional approach expands on the application to more than one dimensional data. Further barriers have been removed by involving tensors factorisation theory as it pushes the applications beyond a two-dimensional decomposition where the factors are the eigentriple components. It nicely moves to the high dimensional domain and imposes some conditions on the relation between the rank of a tensor, constructed from a single channel signal only, and the number of components within that signal. This under research work further enhances the accuracy of singular spectrum analysis by giving more insight into the selection of embedding dimension.

In the domain of recently introduced hypercomplex signal processing one can merge multimodal or multichannel information into one hypercomplex signal and apply the decomposition. This new area however, introduces the notion of augmented statistics into the singular value decomposition of complex signals. Further research will be necessary in order to develop and apply widely linear models (as opposed to linear models) for the cases where any optimisation procedure for estimating the system parameters will be necessary.

The algorithms can always be further tailored for particular applications; where the signals are nonstationary, the time series from different channels are correlated, or the signals are naturally complex (as they are in a transform domain). The applications to biomedical signals and images in this book are indeed limited. There is always a need for accessing the primitive features and fundamental subcomponents of the data for segmentation, detection, separation, prediction, compression, etc. This is often very problematic for single channel (or limited number of channels) data. Therefore, an area for future research may be devoted to incorporating other possible diversities or properties of those subcomponents for their extraction. Here we started with frequency diversity of the constituent sources in introducing an empirical mode decomposition approach for the extraction of narrow band signals. We also relied on some physical properties of the sources of interest in their separation process. Such signals however, may originate from spatially or statistically different sources. Such information can be exploited in some applications too.

Although biomedical signals can be very different in nature, they have much in common; they are usually very noisy. The noise can be an internal noise as the result of body activity and changes in metabolism, or external, as the influences of environment are tremendous. The signals often include some periodic or cyclo-stationary components; a noise-type electromyography signal from muscles contains a quasi-periodic component of cardiac potential or a signal from the brain of an awake person often contains one to four different frequency components. This suggests that the frequency band of interest can help in a more accurate identification of the desired factors within the signal subspaces.

Rather than the above properties, the recent models in singular spectrum analysis can exploit the links and dependencies between the trends. The signals recorded from a human have much in common and upon recognition of these correlations and similarities such methods can be better used.

The major stages of decomposition, grouping, and reconstruction have their natural roots in many physiological processes starting from single cell analysis to colonial theory leading to the amazing topic of genomics. This also emphasises the modelling capability of singular spectrum analysis which can open a new front for modelling biological and physiological phenomena.

Therefore, despite its simplicity, singular spectrum analysis not only has the potential for many applications but also can join other signal processing algorithms during their pre-processing (such as for enhancement of images or denoising the signals) or post-processing (such as for feature detection) stages to form a more powerful hybrid system. As another merit, the future states of complex models can be predicted by means of singular spectrum analysis. This area of research also has room to develop, as there are many external factors which influence a trend. The art of signal processing in recent years has flourished by developing regularised algorithms to take these factors into account and mathematically incorporate them into decomposition or reconstruction formulations.

This book therefore, brings more hope in expanding this simple though very effective concept for many other applications in signal, image, and data processing.

Index